金仓数据库
KingbaseES
PL/SQL编程

张德珍 张 俊 曹志英 杜 胜 冯 玉 **编著**

清華大学出版社
北 京

内 容 简 介

本书基于金仓数据库管理系统 KingbaseES V9 版本,采用"理论+实践"的形式编写。在全面介绍 KingbaseES PL/SQL 基本知识的基础上,详细讲解了 PL/SQL 开发的所有功能模块、方法和技巧,包括 PL/SQL 中的静态 SQL 语句、事务处理、动态 SQL 语句、异常处理、输入与输出、子程序、用户自定义对象、用户自定义聚集函数、程序包、触发器、代码保护、调试、调度与执行等主要内容。

本书以一个简化的在线购物平台数据库为例,将理论知识翔实地融入实践当中,以大量应用实例验证、解读,让读者体验完整的项目实操过程。此外,考虑与 Oracle 及 PostgreSQL 数据库的兼容性,协助数据库开发人员用好国产数据库,提高国产数据库在国民经济各领域的应用范围,本书还作了差异性比较和转换操作。

本书适用于 KingbaseES 数据库开发人员、KingbaseES 数据库管理员等相关数据库从业人员,也可作为大中专院校计算机科学与技术、软件工程等相关专业师生的参考用书,还可作为培训机构的培训教材。

本书封面贴有清华大学出版社防伪标签,无标签者不得销售。
版权所有,侵权必究。举报: 010-62782989,beiqinquan@tup.tsinghua.edu.cn。

图书在版编目(CIP)数据

金仓数据库 KingbaseES PL/SQL 编程/张德珍等编著. —北京:清华大学出版社,2023.8(2024.6 重印)
ISBN 978-7-302-63930-5

Ⅰ.①金… Ⅱ.①张… Ⅲ.①关系数据库系统-程序设计 Ⅳ.①TP311.132.3

中国国家版本馆 CIP 数据核字(2023)第 115956 号

责任编辑:张 玥
封面设计:常雪影
责任校对:胡伟民
责任印制:沈 露

出版发行:清华大学出版社
网　　址:https://www.tup.com.cn,https://www.wqxuetang.com
地　　址:北京清华大学学研大厦 A 座
邮　　编:100084
社 总 机:010-83470000
邮　　购:010-62786544
投稿与读者服务:010-62776969,c-service@tup.tsinghua.edu.cn
质量反馈:010-62772015,zhiliang@tup.tsinghua.edu.cn
课件下载:https://www.tup.com.cn,010-83470236

印 装 者:三河市东方印刷有限公司
经　　销:全国新华书店
开　　本:185mm×260mm　　印　张:20.25　　字　数:510 千字
版　　次:2023 年 9 月第 1 版　　印　次:2024 年 6 月第 2 次印刷
定　　价:75.00 元

产品编号:100502-01

前 言

  金仓数据库 KingbaseES 是北京人大金仓信息技术股份有限公司（简称"人大金仓"）研发的一款面向大规模并发交易处理的企业级关系数据库。它融合了人大金仓在数据库领域几十年产品研发与企业级应用的实践经验，可满足各行业用户多种场景的数据处理需求。KingbaseES 遵循严格的 ACID 特性，结合多核架构的极致性能、行业最高的安全标准、完备的高可用方案，以及可覆盖迁移、开发及运维管理全生命周期的智能便捷工具，可为用户带来极致的使用体验。金仓数据库 KingbaseES 广泛服务于电子政务、能源、金融、电信、教育及交通等 60 余个重点行业和关键领域，累计装机部署超过 100 万套，入选国务院国资委发布的十项国有企业数字技术典型成果。

  PL/SQL 是 KingbaseES 数据库对标准 SQL 语言进行过程化扩展，专门用于各种环境下对 KingbaseES 数据库进行访问和开发的语言，具有高性能、可移植、可扩展、兼容性好、支持面向对象编程等优点。

  本书从 PL/SQL 的基本语法入手，详细讲解了 PL/SQL 开发的所有功能模块、方法和技巧，并以一个简化的在线购物平台数据库为例，将各章节的理论知识翔实地融入实践当中，让读者体验完整的项目实操过程，也为 KingbaseES 开发者提供一个关于 PL/SQL 过程语言的全面、坚实的参考资源。本书既适用于 KingbaseES 数据库开发人员、KingbaseES 数据库管理员等相关数据库从业人员，也可以作为大中专院校计算机科学与技术、软件工程等相关专业师生的参考用书和培训机构的培训教材。

  全书共分 16 章，章节安排以 KingbaseES PL/SQL 的功能特点为主线展开，通过大量系统性应用实例加以验证，内容讲解由浅入深，层次清晰，通俗易懂。第 1 章介绍 KingbaseES PL/SQL 的优点、特征、运行机制和案例数据库；第 2 章介绍 PL/SQL 的程序结构以及涉及的 3 种主要控制语句；第 3 章介绍 PL/SQL 的复合数据类型，包括集合类型和记录类型；第 4 章介绍游标、游标变量、批量处理操作等 PL/SQL 中的静态 SQL 语句及其应用；第 5 章介绍事务处理、自治事务，二者之间的调用关系及使用；第 6 章介绍动态 SQL 语句，包括本地动态 SQL 和 DBMS_SQL 包两种技术，以及各自的编程方法；第 7 章介绍处理 PL/SQL 运行错误的方法，包括定义异常、引发异常和处理异常；第 8 章介绍 PL/SQL 中 3 种常用的输入与输出机制，包括在控制台输入和输出信息，读取数据到内存和写入数据到本地，以及通过网络获取数据；第 9 章介绍 PL/SQL 子程序，包括独立子程序、嵌套子程序以及表函数；第 10 章介绍 PL/SQL 中用户自定义对象的创建、使用，以及与 Oracle 的差异；第 11 章介绍 PL/SQL 中用户自定义聚集函数的创建、使用，以及与 Oracle 的差异；第 12 章介绍程序包的创建和使用；第 13 章介绍 PL/SQL 中 DML 触发器、事件触发器的创建和使用，以及触发器的管理；第 14 章介绍 PL/SQL 中通过 Wrapper 实用程序和 DBMS_DDL 子程序实现代码的加密保护；第 15 章介绍 PL/SQL 的执行跟踪和调试器功能；第 16 章介绍通过 DBMS_JOB 和 DBMS_SCHEDULER 包的自动作业功能函数实现任务调度与执行功能。

本书具有以下特点。

（1）由浅入深地成体系讲解 KingbaseES PL/SQL 的所有功能，弥补了国产数据库在 PL/SQL 过程语言方面的内容缺失；该体系下的丰富内容和组织形式使数据库开发人员更易于理解、掌握相关理论和方法。

（2）注重理论和实践的结合，以一个简化的在线购物平台数据库案例贯穿全书，讲解理论时针对每一个语法或知识点，结合案例设计了大量应用实例验证、解读，使读者能读懂、会使用，提高效率。

（3）考虑从国外数据库到国产数据库的迁移需求，如从 Oracle 及 PostgreSQL 数据库迁移到 KingbaseES，协助数据库开发人员用好国产数据库，提高国产数据库在国民经济各领域的应用范围，书中对相关数据库作了差异性比较，并提供了丰富的转换操作应用实例，使 KingbaseES 能更好地兼容其他数据库。

（4）提供配套的数据库案例和示例源代码。

本书由张德珍、张俊、曹志英、杜胜、冯玉共同编写。全书大纲由杜胜、冯玉拟制，由张德珍执笔，张俊、曹志英参与本书的编写讨论和用户案例库设计，最后由张德珍统稿，杜胜和冯玉参与本书撰写过程的全部讨论，并对本书进行审定，提出大量宝贵的意见和建议。参与本书编写工作的还有大连拓扑伟业科技有限公司的宋盈吉、王兆松、郑昊、朱莹琦，大连海事大学信息科学与技术学院的宋鹏飞、秦一、陶李婷、相龙、刘畅等同学，分别收集整理了各章节相关素材，并设计和验证了相关案例。在编写过程中，作者参阅了人大金仓、甲骨文公司、PostgreSQL 开源数据库、大连拓扑伟业科技有限公司等相关的数据库文档、联机帮助和教学培训成果，也吸取了国内外相关参考书的精髓，对这些作者的贡献表示由衷的感谢。在本书的出版过程中，得到了中国人民大学王珊教授的支持和帮助，还得到了清华大学出版社张玥编辑的大力支持，在此表示诚挚的感谢。

由于作者水平有限，书中难免有不妥和疏漏之处，恳请各位专家、同仁和读者不吝赐教，并与笔者讨论。

作　者
2023 年 5 月于大连

目 录

第 1 章　KingbaseES PL/SQL 概述 1
1.1　PL/SQL 概述 1
1.1.1　PL/SQL 的优点 1
1.1.2　PL/SQL 的特性 2
1.1.3　PL/SQL 的结构 4
1.2　创建与运行 PL/SQL 代码 5
1.2.1　KSQL 6
1.2.2　KStudio 9
1.3　PL/SQL 的运行机制 10
1.3.1　PL/SQL 引擎 10
1.3.2　PL/SQL 单元与编译参数 12
1.4　案例数据库介绍 14
1.4.1　SeaMart 的 E-R 图及其表结构 14
1.4.2　创建表 19
1.5　使用金仓在线帮助文档 23

第 2 章　PL/SQL 程序结构 24
2.1　PL/SQL 块结构 24
2.2　PL/SQL 声明段 27
2.2.1　数据类型 27
2.2.2　变量和常量的声明 28
2.3　基本执行语句 29
2.4　条件选择语句 32
2.4.1　IF 32
2.4.2　CASE 36
2.5　循环语句 38
2.5.1　基本循环语句 39
2.5.2　FOR LOOP 语句 41
2.5.3　WHILE LOOP 语句 43
2.5.4　FOREACH 语句 44

2.6 获取执行状态信息 ·· 45
　　2.6.1 获取结果状态和执行位置信息 ·· 45
　　2.6.2 错误和消息 ·· 47

第 3 章　PL/SQL 的复合数据类型 ··· 49
3.1 集合类型 ·· 49
　　3.1.1 关联数组 ·· 50
　　3.1.2 可变数组 ·· 52
　　3.1.3 嵌套表 ·· 53
　　3.1.4 集合的构造函数 ··· 56
　　3.1.5 集合变量赋值 ·· 57
　　3.1.6 多维集合 ·· 59
　　3.1.7 集合的比较 ··· 60
　　3.1.8 集合方法 ·· 61
3.2 记录类型 ·· 65
　　3.2.1 记录类型概述 ·· 65
　　3.2.2 声明记录类型 ·· 65
　　3.2.3 使用记录类型 ·· 66

第 4 章　PL/SQL 中的静态 SQL 语句 ··· 69
4.1 静态 SQL 语句概述 ·· 69
　　4.1.1 静态 SQL 语句类型 ··· 69
　　4.1.2 PL/SQL 中的 SELECT 语句 ··· 70
　　4.1.3 PL/SQL 中的 DML 语句 ·· 73
4.2 游标 ·· 75
　　4.2.1 游标概念 ·· 75
　　4.2.2 隐式游标 ·· 76
　　4.2.3 声明和定义显式游标 ·· 78
　　4.2.4 打开和关闭显式游标 ·· 79
　　4.2.5 使用显式游标获取数据 ··· 79
　　4.2.6 显式游标查询中的变量 ··· 84
　　4.2.7 当显式游标查询需要列别名时 ·· 86
　　4.2.8 接收参数的显式游标 ·· 87
　　4.2.9 显式游标属性 ·· 90
4.3 游标变量 ··· 93
　　4.3.1 创建游标变量 ·· 94
　　4.3.2 打开和关闭游标变量 ·· 95
　　4.3.3 使用游标变量获取数据 ··· 95
　　4.3.4 为游标变量赋值 ··· 97

		4.3.5 游标变量查询中的变量	98
		4.3.6 游标变量属性	100
		4.3.7 游标变量作为子程序参数	100
	4.4	批量处理	102

第5章 事务处理 … 109

- 5.1 事务处理概述 … 109
- 5.2 事务处理语句 … 109
  - 5.2.1 COMMIT 语句 … 109
  - 5.2.2 ROLLBACK 语句 … 109
  - 5.2.3 SET TRANSACTION 语句 … 110
- 5.3 自治事务 … 111
  - 5.3.1 声明自治事务 … 113
  - 5.3.2 从 SQL 中调用自治函数 … 115

第6章 动态 SQL 语句 … 116

- 6.1 动态 SQL 语句概述 … 116
- 6.2 Native dynamic SQL … 118
  - 6.2.1 EXECUTE IMMEDIATE 语句 … 119
  - 6.2.2 OPEN FOR、FETCH 和 CLOSE 语句 … 124
  - 6.2.3 重复的占位符名称 … 125
- 6.3 DBMS_SQL 包 … 127
  - 6.3.1 DBMS_SQL 包中的常用方法 … 127
  - 6.3.2 DBMS_SQL 包操作流程 … 128
  - 6.3.3 其他常用 DBMS_SQL 方法 … 131
- 6.4 SQL 注入 … 136
  - 6.4.1 SQL 注入技术 … 136
  - 6.4.2 防范 SQL 注入 … 139

第7章 异常处理 … 142

- 7.1 异常处理的概念和术语 … 142
  - 7.1.1 异常种类 … 143
  - 7.1.2 异常处理程序的优点 … 143
- 7.2 定义异常 … 145
  - 7.2.1 系统预定义异常 … 145
  - 7.2.2 用户自定义异常 … 147
  - 7.2.3 重新声明预定义的异常 … 148
- 7.3 引发异常 … 149
  - 7.3.1 显式触发异常 … 149

		7.3.2 异常传播	153
		7.3.3 未处理的异常	160
	7.4	处理异常	161
		7.4.1 处理异常的措施	161
		7.4.2 检索异常信息	162
		7.4.3 异常捕获	165
		7.4.4 获取异常状态信息	168
		7.4.5 检查断言	169

第 8 章 PL/SQL 中的输入与输出 … 171

	8.1	显示信息(DBMS_OUTPUT)	171
		8.1.1 启用 DBMS_OUTPUT	171
		8.1.2 向缓冲区输入信息	172
		8.1.3 从缓冲区读取信息	173
	8.2	文件读写	174
		8.2.1 启动 UTL_FILE	174
		8.2.2 UTL_FILE 方法	174
	8.3	使用基于 Web 的数据(http)	177
		8.3.1 UTL_HTTP 数据类型	177
		8.3.2 UTL_HTTP 方法	178
		8.3.3 http 数据类型	181
		8.3.4 http 方法	183

第 9 章 PL/SQL 子程序 … 187

	9.1	子程序概述	187
		9.1.1 子程序的分类	187
		9.1.2 子程序的优点	187
	9.2	独立子程序	188
		9.2.1 子程序结构	188
		9.2.2 创建函数	189
		9.2.3 创建存储过程	190
		9.2.4 支持的参数	192
		9.2.5 调用与使用	195
		9.2.6 支持的返回值类型	197
	9.3	嵌套子程序	200
		9.3.1 概述	200
		9.3.2 声明和定义	201
		9.3.3 支持的参数	203
		9.3.4 调用与变量	204

9.4 子程序重载 ………………………………………………………………………… 205
9.5 表函数 ……………………………………………………………………………… 206
 9.5.1 结果返回行集合 ………………………………………………………… 206
 9.5.2 结果返回集合数据类型 ………………………………………………… 209

第 10 章 用户自定义对象 …………………………………………………………… 211

10.1 用户自定义对象概述 ……………………………………………………………… 211
10.2 创建用户自定义对象 ……………………………………………………………… 211
 10.2.1 对象类型 ………………………………………………………………… 211
 10.2.2 对象实例 ………………………………………………………………… 215
10.3 在 PL/SQL 中使用自定义对象 …………………………………………………… 216
 10.3.1 定义对象 ………………………………………………………………… 216
 10.3.2 初始化对象 ……………………………………………………………… 216
 10.3.3 调用构造函数 …………………………………………………………… 217
 10.3.4 调用 MEMBER 方法和 STATIC 方法 …………………………………… 217
 10.3.5 对象表的 DML 操作 …………………………………………………… 218
10.4 与 Oracle 数据库中对象类型的差异 ……………………………………………… 219

第 11 章 用户自定义聚集函数 ……………………………………………………… 222

11.1 用户自定义聚集函数概述 ………………………………………………………… 222
 11.1.1 聚集函数 ………………………………………………………………… 222
 11.1.2 创建用户自定义聚集函数 ……………………………………………… 223
11.2 用户自定义聚集函数的运用 ……………………………………………………… 224
 11.2.1 场景数据 ………………………………………………………………… 224
 11.2.2 创建用户自定义聚集函数 ……………………………………………… 225
 11.2.3 用户自定义聚集函数的使用 …………………………………………… 226
 11.2.4 查看用户自定义聚集函数信息 ………………………………………… 227
11.3 KingbaseES 与 Oracle 中创建聚集函数的差异 ………………………………… 228

第 12 章 程序包 ……………………………………………………………………… 232

12.1 程序包概述 ………………………………………………………………………… 232
 12.1.1 包的概念 ………………………………………………………………… 232
 12.1.2 包的优点 ………………………………………………………………… 232
 12.1.3 系统内置包 ……………………………………………………………… 233
12.2 创建程序包 ………………………………………………………………………… 233
 12.2.1 包的组成 ………………………………………………………………… 233
 12.2.2 包的创建 ………………………………………………………………… 235
12.3 程序包的使用 ……………………………………………………………………… 239
 12.3.1 包元素的调用规则 ……………………………………………………… 239

　　　　12.3.2　包数据 ………………………………………………… 240
　　　　12.3.3　包游标 ………………………………………………… 240
　　　　12.3.4　查看程序包信息 ……………………………………… 242

第 13 章　触发器 ……………………………………………………… 247
13.1　触发器简介 …………………………………………………… 247
　　　　13.1.1　触发器的概念 ………………………………………… 247
　　　　13.1.2　触发器的作用 ………………………………………… 248
　　　　13.1.3　触发器的种类 ………………………………………… 248
13.2　DML 触发器 ………………………………………………… 248
　　　　13.2.1　DML 触发器的用途 …………………………………… 248
　　　　13.2.2　创建 DML 触发器 ……………………………………… 248
　　　　13.2.3　触发器体 ……………………………………………… 250
　　　　13.2.4　INSTEAD OF 触发器 ………………………………… 253
　　　　13.2.5　触发器触发的顺序 …………………………………… 254
13.3　事件触发器 …………………………………………………… 254
　　　　13.3.1　事件触发器概述 ……………………………………… 254
　　　　13.3.2　创建事件触发器 ……………………………………… 255
13.4　触发器设计注意事项 ………………………………………… 255
13.5　触发器管理 …………………………………………………… 256
　　　　13.5.1　禁用与启用触发器 …………………………………… 256
　　　　13.5.2　修改、重编译与删除触发器 ………………………… 257
　　　　13.5.3　触发器信息查询 ……………………………………… 257

第 14 章　PL/SQL 的代码加密 …………………………………… 259
14.1　PL/SQL 代码加密概述 ……………………………………… 259
14.2　Wrapper …………………………………………………… 260
　　　　14.2.1　使用 PL/SQL Wrapper 实用程序 …………………… 260
　　　　14.2.2　PL/SQL Wrapper 实用程序的输入与输出文件 …… 260
　　　　14.2.3　PL/SQL Wrapper 加密的优点和局限性 …………… 261
　　　　14.2.4　示例 …………………………………………………… 261
14.3　DBMS_DDL 包的使用 ……………………………………… 264
　　　　14.3.1　使用 DBMS_DDL 子程序 ……………………………… 264
　　　　14.3.2　DBMS_DDL 加密的局限性 …………………………… 264
　　　　14.3.3　示例 …………………………………………………… 264

第 15 章　PL/SQL 的调试 ………………………………………… 266
15.1　PL/SQL 的执行跟踪 ………………………………………… 266
　　　　15.1.1　DBMS_UTILITY ……………………………………… 267

		15.1.2 性能监控 ··· 271
15.2	PL/SQL 调试器 ··· 277
	15.2.1	函数/存储过程调试 ··· 277
	15.2.2	触发器调试 ··· 283
	15.2.3	程序包调试 ··· 285

第 16 章 PL/SQL 任务的调度与执行 ··· 290

16.1 使用 DBMS_JOB 包管理任务 ··· 290
    16.1.1 任务的创建 ··· 291
    16.1.2 任务的执行 ··· 296
    16.1.3 任务的删除 ··· 297

16.2 使用 DBMS_SCHEDULER 包管理任务 ··· 299
    16.2.1 任务的创建 ··· 299
    16.2.2 任务的执行 ··· 304
    16.2.3 任务的删除 ··· 304

16.3 使用 KStudio 管理任务 ··· 307
    16.3.1 任务的创建 ··· 307
    16.3.2 任务的执行 ··· 308
    16.3.3 任务的删除 ··· 309

参考文献 ·· 312

第 1 章

KingbaseES PL/SQL 概述

PL/SQL(Procedural Language extensions to SQL)是 KingbaseES 数据库对标准 SQL 进行过程化扩展,专门用于各种环境下对 KingbaseES 数据库进行访问和开发的语言。本章主要介绍以下内容。
- PL/SQL 概述。
- 创建与运行 PL/SQL 代码。
- PL/SQL 运行机制。
- 案例数据库介绍。
- 使用金仓在线帮助文档。

 ## 1.1 PL/SQL 概述

标准的 SQL 是高度的非过程化语言,可以简单、高效、快捷地定义各种数据库对象,操作数据库中的数据,但不具有过程化编程语言中的流程控制操作、异常处理操作、事件处理等特点,并且大部分语句的执行与其前后语句无关,无法实现复杂的业务处理逻辑。因此,KingbaseES 数据库对标准的 SQL 进行了过程化扩展,在原来 SQL 非过程化的基础上扩充并赋予了其编程语言的特点,增加了过程化的元素,包括变量、条件选择、循环控制、错误处理、集合、子程序、包等操作,可以实现对象类型、存储过程、触发器等复杂的功能,成为具备计算功能的程序语言,是一种用于数据库系统的可载入过程语言。

1.1.1 PL/SQL 的优点

1. 与 SQL 紧密集成

PL/SQL 与广泛使用的数据库操作语言 SQL 紧密集成。举例如下:

(1) PL/SQL 允许使用所有 SQL 数据操作、游标控制和事务控制语句,以及所有 SQL 函数、运算符和伪列。

(2) PL/SQL 完全支持 SQL 数据类型,无须在 PL/SQL 和 SQL 数据类型之间进行转换。

(3) PL/SQL 允许运行 SQL 查询,并一次处理查询结果集中的一行或多行。

(4) PL/SQL 函数可以在 SQL SELECT 语句的 WITH 子句中声明和定义。

(5) PL/SQL 支持静态和动态 SQL,动态 SQL 使应用程序更加灵活和通用。

2. 高性能

PL/SQL 允许将语句块发送到数据库,从而显著减少应用程序和数据库之间的流量。功能如下。

(1) 绑定变量。

在 PL/SQL 代码中直接嵌入 SQL 的 INSERT、UPDATE、DELETE、MERGE 或 SELECT 语句时,PL/SQL 编译器会将 WHERE 和 VALUES 子句中的变量转换为绑定变量。KingbaseES 数据库可以在每次运行相同的代码时重用这些 SQL 语句,从而提高性能。使用动态 SQL 时,PL/SQL 不会自动创建绑定变量,但可以通过显式指定将其与动态 SQL 一起使用。

(2) 子程序。

PL/SQL 子程序以明文或加密的形式存储。每个 session 在第一次调用子程序时都需要进行一次编译,且保存编译结果。

3. 高生产力

PL/SQL 有许多节省设计和调试时间的特性,并且在所有环境中都一样。PL/SQL 允许编写紧凑的代码操作数据,可以查询、转换和更新数据库中的数据。

4. 可移植性

PL/SQL 是用于数据库开发的可移植标准语言。可以在运行数据库的任何操作系统和平台上运行 PL/SQL 应用程序。

5. 可扩展性

PL/SQL 存储子程序通过将应用程序处理集中在数据库服务器上来提高可扩展性。

6. 可管理性

PL/SQL 存储的子程序提高了可管理性,任何应用程序都可以使用子程序、更改子程序,而不影响调用它们的应用程序。

7. 支持面向对象编程

PL/SQL 允许定义可用于面向对象设计的对象类型,支持带有"抽象数据类型"的面向对象编程。

1.1.2 PL/SQL 的特性

PL/SQL 结合了 SQL 的数据操作能力和过程语言的处理能力。使用 SQL 解决问题时,也可以从 PL/SQL 程序执行 SQL 语句,而无须学习新的应用程序接口(API)。与其他过程编程语言一样,PL/SQL 允许声明常量和变量、控制程序流、定义子程序和捕获运行时错误。可以将复杂的问题分解为易于理解的子程序,并且在不同的应用程序中重复使用这些子程序。

1. 错误处理

PL/SQL 使检测和处理错误变得容易。发生错误时,PL/SQL 会引发异常,正常执行停止,并将控制转移到 PL/SQL 块的异常处理部分。

2. 块

PL/SQL 源程序的基本单元是块，由关键字 DECLARE、BEGIN、EXCEPTION 和 END 定义。这些关键字将块分为声明部分、可执行部分和异常处理部分，块可以嵌套。

3. 变量和常量

PL/SQL 允许声明变量和常量，在可以使用表达式的任何地方使用它们。

4. 子程序

PL/SQL 子程序是一个命名的 PL/SQL 块，可以重复调用。PL/SQL 有两种类型的子程序，即过程和函数，函数可以具有返回值。

5. 包

包是一个模式对象，可以对逻辑相关的 PL/SQL 类型、变量、常量、子程序、游标和异常进行分组。一个包被存储在数据库中，许多应用程序可以访问并使用它的内容。

6. 触发器

触发器是一个命名的 PL/SQL 单元，它存储在数据库中，并响应数据库中发生的事件而运行。可以指定触发器是在事件之前还是之后触发，以及触发器是针对每个事件还是针对受事件影响的每一行运行。

7. 输入和输出

大多数 PL/SQL 输入和输出（I/O）都是通过在数据库表中存储数据或查询这些表的 SQL 语句完成的。所有其他 PL/SQL 的 I/O 都是使用 KingbaseES 数据库提供的 PL/SQL 包（dbms_output,UTL_FILE）完成的。

8. 数据抽象

通过数据抽象可以处理数据的基本属性，而无须过多关注细节。

（1）游标。

游标是指向私有 SQL 区域的指针，该区域存储有关处理特定 SQL 语句或 PL/SQL SELECT INTO 语句的信息。可以使用游标一次检索一行的结果集，也可以使用游标属性获取有关游标状态的信息。

（2）复合变量。

复合变量具有内部组件，可以单独访问这些组件，也可以将整个复合变量作为参数传递给子程序。PL/SQL 有两种复合变量：集合和记录。在集合中，内部组件始终具有相同的数据类型，称为元素，可以通过其唯一索引访问每个元素。在记录中，内部组件可以是不同的数据类型，称为字段，通过名称可以访问每个字段。

（3）使用％ROWTYPE 属性。

％ROWTYPE 属性允许声明数据库表或视图的完整或部分行的 record 变量。对于整行或部分行的每一列，record 变量都有一个具有相同名称和数据类型的字段。如果行的结构发生变化，那么记录的结构也会相应地发生变化。

（4）使用％TYPE 属性。

％TYPE 属性允许在不知道数据类型的情况下声明与先前声明的变量或列具有相同数据类型的数据对象。如果被引用的对象声明发生变化，那么引用对象的声明也会相应改变。

(5) 抽象数据类型

抽象数据类型（ADT）由数据结构和操作数据的子程序组成。构成数据结构的变量称为属性,操纵属性的子程序称为方法。ADT 存储在数据库中,ADT 的实例可以存储在表中,并用作 PL/SQL 变量。ADT 可以将大型系统分成可重用的逻辑组件,以降低复杂性。

9. 控制语句

控制语句是 PL/SQL 语言对 SQL 语言最重要的扩展。PL/SQL 有三类控制语句:条件选择语句、循环语句、顺序控制语句。

10. 一次处理查询结果集的一行

PL/SQL 允许发出 SQL 查询,并一次处理结果集的一行。既可以使用基本循环,也可以通过使用单独的语句来运行查询,检索结果,实现精确控制。

1.1.3　PL/SQL 的结构

PL/SQL 程序以语句块(block)为单位,每个语句块由声明部分、执行语句和异常处理部分组成,定义如下：

```
[ <<label>> ]                --语句块标识(可选)
[ DECLARE                    --声明部分(可选)
     Declarations ]
BEGIN                        --执行部分(必需)
     Statements              --语句(可以使用声明部分的元素)
  [EXCEPTION                 --异常处理部分(可选)
     Statements ]            --异常处理操作(处理执行部分抛出的异常)
END [ label ];
```

关于语句块更为详细的描述,将在第 2 章讲述。基本的代码单元是语句块,语句块是可以嵌套的,即嵌套另一个 PL/SQL 语句块,每个语句块完成特定的功能。

PL/SQL 语句块按照是否在数据库中存储分为两种:匿名块和命名块。

1. 匿名块

匿名块指的是没有名称、不在数据库中存储,不能被其他程序调用,但可以调用其他子程序的 PL/SQL 程序。PL/SQL 匿名块每次执行时都需要程序的源代码,先编译后执行。因为匿名块可以有它自己的声明和异常处理部分,所以开发人员经常会在一个大的程序中嵌套一些匿名块,以提供一系列的标识符和异常处理。

定义 PL/SQL 匿名块的通用语法与块结构一致,声明和异常部分是可选的,执行部分是必须存在的,在 KingbaseES 中可以没有任何语句。

2. 命名块

命名块通常指的是存储在数据库中可以多次执行的 PL/SQL 程序,包括函数、存储过程、包、触发器等。它们编译后放在数据库服务器中,由应用程序或系统在特定条件下调用执行。大多数编写的代码还是以命名块的形式出现。

命名块与匿名块的不同之处在于,命名块是存储在数据库中可以多次执行的 PL/SQL 程序。

示例 1.1：PL/SQL 匿名块程序。

功能描述：查询 adminaddrs 表中"中华人民共和国"的编码 id。

程序代码如下。

```
DECLARE
  SAL_TABLE  VARCHAR2;
  s_ADDRNAME VARCHAR2;
BEGIN
  s_ADDRNAME:= '中华人民共和国';
  SELECT ADDRID INTO SAL_TABLE FROM ADMINADDRS
  WHERE ADDRNAME= s_ADDRNAME;
  RAISE NOTICE 'SAL_TABLE=%',SAL_TABLE ;

  EXCEPTION WHEN NO_DATA_FOUND
    THEN RAISE NOTICE '%','没有查到相关数据。';
END;
```

示例 1.2：PL/SQL 命名块程序。

功能描述：将查找地址编码 id 的操作封装成 searchaddID 过程,然后可反复调用该命名块。

程序代码如下。

```
CREATE OR REPLACE PROCEDURE searchaddID (
  IN s_ADDRNAME VARCHAR2
) AS
DECLARE SAL_TABLE VARCHAR2;
BEGIN
  SELECT ADDRID
  INTO SAL_TABLE
  FROM ADMINADDRS
  WHERE ADDRNAME= s_ADDRNAME;
END;
```

调用该过程的代码如下。

```
CALL searchaddID('中华人民共和国');
```

程序运行结果如下。

```
SQL 错误 [P0002]：错误：查询没有返回记录
```

1.2　创建与运行 PL/SQL 代码

"工欲善其事,必先利其器。"选择合适的开发工具并熟练应用,在 PL/SQL 学习与开发中可以起到事半功倍效果。本节将介绍以下两种常用的 PL/SQL 开发工具及其使用方法。

(1) KSQL。

(2) KStudio。

1.2.1 KSQL

KSQL 作为终端工具,用于与金仓数据库交互。可以通过文件输入或命令行输入方式执行 SQL 语句或 PL/SQL 程序。KSQL 是金仓数据库的一个组件,安装数据库时默认安装。

KSQL 可以在任何金仓数据库运行的操作系统平台上使用,使用方法基本相同,本节中的示例基于 Linux 环境,如图 1.1 所示。

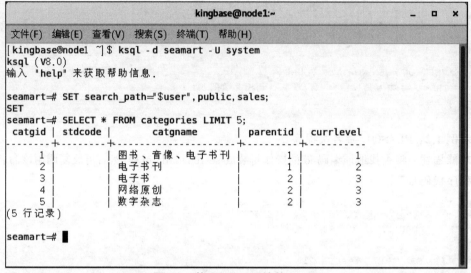

图 1.1 基于 Linux 环境的 KSQL

1. 连接目标数据库

连接到数据库,需要知道 4 个参数,即目标数据库的名称、主机名、端口号以及用于连接数据库的用户名。告知 KSQL 这些参数,可以分别通过命令行选项-d、-h、-p 和-U 完成。如果发现一个参数不属于任何选项,它将被解释为数据库名称(如果已经给出数据库名称,就解释为用户名)。这些选项并非都是必需的,它们都有可用的默认值。

(1) 如果忽略主机名,KSQL 将连接到本地主机的服务器,或通过 TCP/IP 连接到主机上的 localhost。

(2) 数据库服务器使用相同的默认值,大多数情况下不必指定端口。

(3) 默认的用户名和数据库名与操作系统用户名相同。

当默认值不符合实际时,可以将环境变量 KINGBASE_DATABASE、KINGBASE_HOST、KINGBASE_PORT、KINGBASE_USER 设置为适当值。使用 TCP/IP 方式登录数据库有以下两种形式。

在客户端指定 IP 或主机名登录数据库。程序代码如下。

```
$ ksql -h 127.0.0.1 -p 54321 -d seamart -U system
```

使用连接串方式登录数据库。程序代码如下。

```
$ ksql kingbase://system@127.0.0.1:54321/seamart
$ ksql 'hostaddr=127.0.0.1 port=54321 user=system dbname=seamart'
```

正常登录后,KSQL 会提供一个提示符,该提示符是当前连接到的数据库名称后面跟字符串=#。

2. 执行 PL/SQL

(1) 交互方式执行 PL/SQL。

成功连接数据库后,可以通过键入各种 PL/SQL 命令实现有关功能。默认的 KSQL 命令结束符是分号,当碰到代表命令终结的分号时,输入的命令会被发送给服务器。PL/SQL 语句块一行输入的结束并不代表命令的完结,因此,需要使用 SET SQLTERM 命令重新设置命令结束标志。例如,下面的例子使用"/"标识命令的结束。

如果命令执行后不产生错误,那么命令的结果将会显示在屏幕上,例如:

- 使用 system 用户登录到 seamart 数据库。
- 使用 PL/SQL 查询 categorys 表的 catgname 为"手机"的记录。程序代码如下。

```
$ ksql -U system -d seamart
seamart=# \set SQLTERM /
seamart=# DECLARE
seamart-#    s_catgname VARCHAR2;
seamart-# BEGIN
seamart-#    s_catgname:='手机';
seamart-#    SELECT * FROM sales.categories
seamart-#    WHERE catgname=s_catgname;
seamart-# END;
seamart-# /
```

程序运行结果如下。

```
 catgid | stdcode | catgname | parentid | currlevel |
--------+---------+----------+----------+-----------+
     2|    手机 |        1|        2|
     3|    手机 |        2|        3|
     4|    手机 |        2|        3|
    74|    手机 |       74|        1|
    75|    手机 |       74|        2|
    76|    手机 |       75|        3|
  2075|    手机 |     2074|        1|
```

(2) 非交互方式执行 PL/SQL(运行脚本)。

几乎所有可以在 KSQL 终端环境中交互运行的语句都可以存成脚本文件,以便重复执行。

在操作系统当前目录中创建一个脚本文件 test.sql,程序代码如下。

```
\set SQLTERM /
DECLARE
```

```
    s_catgname VARCHAR2;
BEGIN
  s_catgname:='手机';
    SELECT * FROM sales.categories
  WHERE catgname=s_catgname;
END;
/
```

\i 命令可以从文件中读取输入,并将它作为键盘输入的命令来执行,在 KSQL 中输入如下语句执行 test.sql 文件。

```
seamart=# \i test.sql
```

程序运行结果如下。

```
 catgid | stdcode | catgname | parentid | currlevel |
--------+---------+----------+----------+-----------+
      2 |         | 手机     |        1 |         2 |
      3 |         | 手机     |        2 |         3 |
      4 |         | 手机     |        2 |         3 |
     74 |         | 手机     |       74 |         1 |
     75 |         | 手机     |       74 |         2 |
     76 |         | 手机     |       75 |         3 |
   2075 |         | 手机     |     2074 |         1 |
(7 行记录)
ANONYMOUS BLOCK
```

若要在启动 KSQL 时运行脚本,使用"-f"选项从文件中读取命令,而不是采用标准输入。程序代码如下。

```
$ ksql -d seamart -U system -f test.sql
```

程序运行结果如下。

```
 catgid | stdcode | catgname | parentid | currlevel |
--------+---------+----------+----------+-----------+
      2 |         | 手机     |        1 |         2 |
      3 |         | 手机     |        2 |         3 |
      4 |         | 手机     |        2 |         3 |
     74 |         | 手机     |       74 |         1 |
     75 |         | 手机     |       74 |         2 |
     76 |         | 手机     |       75 |         3 |
   2075 |         | 手机     |     2074 |         1 |
(7 行记录)
ANONYMOUS BLOCK
```

3. 退出 KSQL

输入\q 即可退出 KSQL 命令行。

1.2.2 KStudio

数据库开发管理工具 KStudio 是人大金仓提供的连接金仓数据库的图形化客户端工具,它不仅用于程序开发人员开发数据库项目,还为 DBA 提供了丰富的运维功能。相较命令行工具 KSQL,图形化工具 KStudio 具有如下特点。

(1) 图形化操作方便开发人员编写和调试 SQL 及 PL/SQL 代码。

(2) 对象展示直观,适合初学者快速上手,使用数据库。

(3) 操作简单,省去学习烦琐的命令。

1. 连接数据库

(1) 打开数据库开发管理工具。

(2) 单击"连接"按钮。

- 在"注册新实例"窗口中输入 KES 服务器主机地址、端口、数据库名称、用户名、密码,显示名称可以自定义设置,如图 1.2 所示。
- 单击"检查"进行连接测试。
- 单击"确定"保存连接,并自动登录 KES。

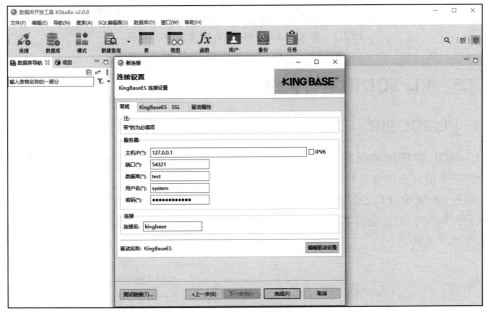

图 1.2　注册新实例窗口

2. 编写与运行 PL/SQL 程序

打开 KStudio,连接数据库后,选中对应的数据库对象,单击"新建查询"按钮,可以开始编写 PL/SQL 程序。输入程序后,单击"执行"按钮或按 Ctrl+Enter 键编译、执行该程序,如果存在编译错误,系统将在"结果"标签提示错误信息,如图 1.3 所示。

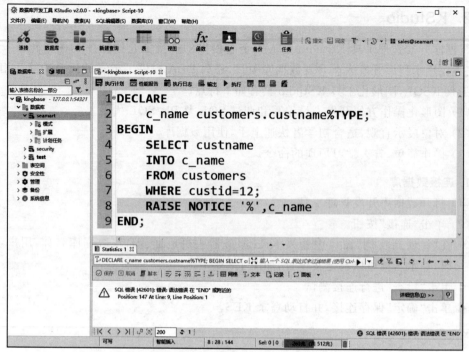

图 1.3 在 KStudio 中编写 PL/SQL 程序并运行

1.3 PL/SQL 的运行机制

1.3.1 PL/SQL 引擎

PL/SQL 程序的编译与执行是通过 PL/SQL 引擎来完成的。KingbaseES 中的 PL/SQL 引擎当前只支持安装到数据库服务器端。

在任一环境中,PL/SQL 引擎都接受任何有效的 PL/SQL 单元作为输入。PL/SQL 引擎执行程序语句,但是会将其中的 SQL 语句发送到数据库中的 SQL 引擎去执行,再获取其结果,如图 1.4 所示。

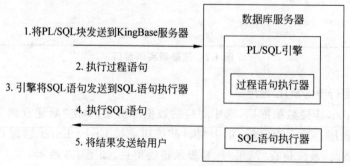

图 1.4 服务器端的 PL/SQL 引擎

当应用程序开发工具处理 PL/SQL 单元时,会将它们传递给其本地 PL/SQL 引擎。如果 PL/SQL 单元不包含 SQL 语句,则客户端引擎处理整个 PL/SQL 单元。如果应用程序开发工具需要条件和迭代控制,PL/SQL 单元将会非常有用。具体如图 1.5 所示。

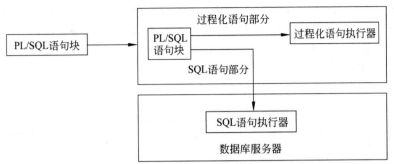

图 1.5　客户端的 PL/SQL 引擎

通常,如果直接利用 SQL 语句对数据库进行操作,各种相关数据的值在代码中以常量形式指定。而在 PL/SQL 中,可以通过变量动态指定各种相关量的值,从而实现对数据库的动态操作。

示例 1.3:在 PL/SQL 语句中执行 SQL 语句。

功能描述:本例中,一个 PL/SQL 匿名块声明了 4 个 PL/SQL 变量,并在静态 SQL 语句 INSERT、UPDATE、SELECT 中使用它们。

程序代码如下。

```
DECLARE
  shop_id shopstores.shopid %TYPE := 2;
  shop_name shopstores.shopname %TYPE :='xm';
  shop_comprgrading shopstores.comprgrading %TYPE := 8.00;
  shop_name1 shopstores.shopname %TYPE;
BEGIN
  IF NOT EXISTS(SELECT shopid FROM sales.shopstores WHERE shopid = shop_id) THEN
  INSERT INTO sales.shopstores VALUES (shop_id , shop_name, NULL, NULL, NULL, NULL, shop_comprgrading);
    RAISE NOTICE '插入一条记录';
  ELSE
  UPDATE shopstores SET shopname =  shop_name, comprgrading = shop_comprgrading
  WHERE shopid = shop_id ;
    RAISE NOTICE '更新该记录';
  END IF;

  SELECT shopname INTO shop_name1 FROM shopstores WHERE shopid = shop_id;
  RAISE NOTICE 'shop_name1:',shop_name1;
END;
```

在该程序中,SELECT、INSERT 和 UPDATE 语句是非过程化的 SQL 语言,完成对数据库的操作;而变量的声明、IF 语句的逻辑判断则是过程化语言的应用。

PL/SQL 程序中引入了变量、控制结构、函数、过程等一系列过程化结构,为更为复杂的数据库应用程序开发提供了可能。

1.3.2 PL/SQL 单元与编译参数

PL/SQL 单元包括匿名块（PL/SQL anonymous block）、函数（FUNCTION）、包（PACKAGE）、包体（PACKAGE BODY）、过程（PROCEDURE）、触发器（TRIGGER）、类型（TYPE）、类型体（TYPE BODY）。PL/SQL 单元受 PL/SQL 编译参数的影响，编译参数为数据库初始化参数。

编译参数有如下几类，如表 1.1 所示。

表 1.1 PL/SQL 编译参数列表

编译参数	描 述
plsql.variable_conflict	变量名和列名冲突时，如何处理： error　　　　　--报错 use_variable　　--解析成变量名（默认） use_column　　--解析成列名
plsql.check_asserts	是否执行 assert 检查，默认值为 true
ora_open_cursors	允许打开的游标数，最大值为 65535，最小值为 0，默认值为 300
plsql.compile_checks	用于控制 PL/SQL 编译时是否进行对象有效性检查。默认值为 false

1. plsql.variable_conflict

在一个 PL/SQL 语句块中的 SQL 语句可以引用块中的变量和参数，当变量名、参数名与表的列名重名时，这条语句就会产生歧义。要解决变量名的冲突问题，则需要限定引用的变量，或者配置变量冲突时的策略。

示例 1.4：依赖于 VARIABLE_CONFLICT 设置的函数。

示例函数中含有 2 个参数，shopid 和 shopname，而 shopstores 表中也含有同名的 2 列。由于设置了编译参数 VARIABLE_CONFLICT 为 use_variable，所以在 UPDATE 命令中 shopname 和 shopid 引用的是函数中的变量，而非 shopstores 表中的列。如果需要引用表中的列，则需要在相应名称前添加表名加以限定，如 WHERE 子句中的 shopstores.shopid。

如果要在系统范围内更改这一行为，需修改配置参数 plsql.variable_conflict，设置为 error、use_variable、use_column 之一。

还可以通过在过程开头插入以下的特殊命令之一单独设置行为，代码如下。

```
/*特殊命令
#variable_conflict error
#variable_conflict use_variable
#variable_conflict use_column
*/
CREATE OR REPLACE PROCEDURE update_shopstores (IN shopid INT, IN shopname character varying)
AS
#variable_conflict use_variable
```

```
BEGIN
  UPDATE shopstores
  SET shopname = shopname
  WHERE shopstores.shopid = shopid ;
END;
```

示例 1.5：不依赖于 VARIABLE_CONFLICT 设置的函数。
程序代码如下。

```
CREATE OR REPLACE PROCEDURE update_shopstores (IN shopid INT, IN shopname
character varying) AS
BEGIN
  UPDATE shopstores
  SET shopname = shopstores.shopname
  WHERE shopstores.shopid = shopid ;
END;
```

2. plsql.check_asserts

assert 断言语句主要用于调试，以便检查 PL/SQL 程序设计是否与预期一致。assert 语句的语法格式如下。

```
assert condition [, message];
```

在该语句中，condition 是一个布尔表达式，预期总是返回 true。如果 condition 结果为 true，则 assert 语句不执行任何操作；如果结果为 false 或 null，则会引发 assert_failure 异常。

通过配置参数 plsql.check_asserts 可以启用或者禁用断言测试。该参数接受布尔值，且默认为 on。如果为 off，则 assert 语句什么也不做。

3. plsql.compile_checks

在 PL/SQL 中，创建一个引用了不存在对象的语句块，该语句块可以正常创建，在被引用的对象补充创建之后，语句块即可正常调用；如果调用时被引用的对象仍不存在，则会抛出异常。

如果希望在创建语句块时就对引用对象的有效性进行检查，则可以通过设置 plsql.compile_checks 参数为 true 编译检查功能的启用。

示例 1.6：关闭编译检查功能，创建引用不存在对象的语句块。
程序代码如下。

```
SET plsql.compile_checks TO off;       --关闭编译检查功能
DROP TABLE IF EXISTS t1;               --删除 t1 表

CREATE OR REPLACE PROCEDURE proc AS
  i int;
BEGIN
  SELECT 1 INTO i FROM t1;             --删除 t1 后,在过程中又引用 t1
END;
```

该过程可以正常创建,调用时会抛出异常,程序代码如下。

```
CALL proc();
SQL 错误 [42P01]: 错误: 关系 "t1" 不存在
```

示例 1.7:开启编译检查功能,创建引用不存在对象的语句块。
程序代码如下。

```
SET plsql.compile_checks TO on;      --开启编译检查功能
DROP TABLE IF EXISTS t1;             --删除 t1 表
CREATE OR REPLACE PROCEDURE proc AS
  i int;
BEGIN
  SELECT 1 INTO i FROM t1;           --删除 t1 后,在过程中又引用 t1
END;
```

此时,创建过程会提示相关引用对象不存在,程序代码如下。

```
NOTICE:关系 "t1" 不存在
```

1.4 案例数据库介绍

本书以一个简化的电商管理系统数据库(SeaMart)为例,后续各章节中所有示例程序都基于该数据库的相关实体及其联系。该系统涉及的业务逻辑的语义描述如下。

(1)一位顾客可以从一个店铺订购不同商品,产生一个订单;一个店铺可以接受多位顾客订购不同商品,产生多个订单。

(2)每个订单由多个供应(商品及其所在店铺)组成,每个供应仅存在于一个订单中。

(3)每个供应包含多个店铺,每个店铺只属于一个供应。

(4)每个供应包含多种产品,每个产品只属于一个供应。

(5)每个商品仅属于一种商品类别,每个商品类别包含多种商品。

(6)一种商品类别包含若干子类,一种子类只从属于一种父级商品类别。

(7)一个订单只有一个收货地址,一个收货地址可以接收多个订单。

(8)一个订单只有一个发货地址,一个发货地址可以承揽多个订单。

(9)一个行政地址包括若干子行政地址,一个子行政地址只从属于一个父级行政地址。

1.4.1 SeaMart 的 E-R 图及其表结构

根据上述语义描述构建 SeaMart 的 E-R 图,如图 1.6 所示。

识别该 E-R 图,并转化为逻辑结构设计,包括 9 个表,分别为店铺表(Shopstores)、商品表(Goods)、店铺商品供应表(Supply)、顾客表(Customers)、商品评论表(Comments)、订单表(Orders)、订单明细表(Lineitems)、商品分类表(Categories)、行政地址表(Adminaddrs)。各表的结构及其约束如表 1.2~表 1.10 所示。

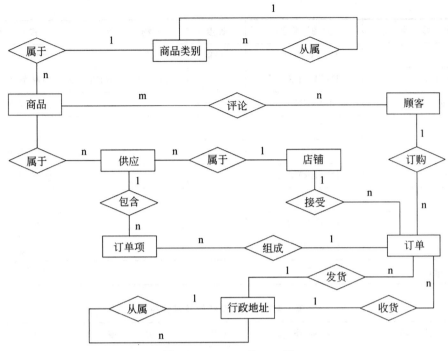

图 1.6 SeaMart 的 E-R 图

表 1.2 Shopstores 表结构及其约束

字 段 名	数 据 类 型	长度	约 束	说 明
shopid	INTEGER		PRIMARY KEY	店铺编号
shopname	VARCHAR	128		店铺名称
shopurl	VARCHAR	256		店铺网址
custgrading	NUMERIC			用户评分
delygrading	NUMERIC			物流评分
servgrading	NUMERIC			服务评分
comprgrading	NUMERIC			综合评分

表 1.3 Goods 表结构及其约束

字 段 名	数 据 类 型	长度	约 束	说 明
goodid	INTEGER		PRIMARY KEY	商品编号
goodname	VARCHAR	128		商品名称
mfrs	VARCHAR	256		生产厂家
brand	VARCHAR	20		商品品牌
model	VARCHAR	50		商品型号
grossweight	REAL			商品毛重

续表

字 段 名	数 据 类 型	长度	约 束	说 明
poo	VARCHAR	128		商品产地
catgid	INTEGER		FOREIGN KEY	商品分类编码
price	MONEY			商品定价
dop	DATE			生产日期
doe	DATE			失效日期
shelflife	INTERVAL			保质期
features	JSON			商品特征
size	POLYGON			商品尺寸
image	BINARY			商品图片
instmanual	TEXT			使用说明

表 1.4　Supply 表结构及其约束

字 段 名	数 据 类 型	长度	约 束	说 明
shopid	INTEGER		PRIMARY KEY/ FOREIGN KEY	店铺编号
goodid	INTEGER		PRIMARY KEY/ FOREIGN KEY	商品编号
totlwhamt	INTEGER			库存总数量
price	MONEY			商品定价
discount	NUMERIC			折扣
status	BOOLEAN			商品状态
url	VARCHAR	512		商品网址
homepage	XML			商品网页转换而来的 XML 数据

表 1.5　Customers 表结构及其约束

字 段 名	数 据 类 型	长度	约 束	说 明
custid	INTEGER		PRIMARY KEY	顾客编号
custidname	VARCHAR	20		顾客姓名
gender	GENDERENUM	256	CHECK (gender IN ('男','女'))	性别
dob	DATE	20		生日
email	VARCHAR	128		电子邮件
ipaddress	INET			用户常用的 IP 地址

续表

字 段 名	数 据 类 型	长度	约 束	说 明
mobile	CHAR	11		手机号码
address	VARCHAR	128		详细地址
mi	INT8RANGE			月收入
ebg	CHAR	4	CHECK（ebg IN('小学','初中','高中','中专','大专','本科','硕士','博士','其他'))	教育程度
prof	CHAR	50		所在行业

表 1.6 Comments 表结构及其约束

字 段 名	数 据 类 型	长度	约 束	说 明
orderid	INTEGER		PRIMARY KEY	订单编号
custid	INTEGER		PRIMARY KEY	顾客编号
goodid	INTEGER		PRIMARY KEY	商品编号
stars	INTEGER			星级
remark	VARCHAR	512		评论
feeling	CHAR	4		情感倾向
submtime	TIMESTAMP WITH TIME ZONE			提交时间

表 1.7 Orders 表结构及其约束

字 段 名	数 据 类 型	长度	约 束	说 明
orderid	INTEGER		PRIMARY KEY	订单编号
custid	INTEGER		FOREIGN KEY	顾客编号
shopid	INTEGER		FOREIGN KEY	店铺编号
submtime	TIMESTAMP WITH TIME ZONE			订单提交时间
ipaddr	INET			用户提交订单时的IP地址
totlbal	MONEY			商品总金额
carriage	MONEY			运费
discamt	MONEY			折扣金额
finlbal	MONEY			最终金额
paytime	TIMESTAMP WITH TIME ZONE			付款时间
paymethod	CHAR	10		付款方式

续表

字 段 名	数 据 类 型	长度	约 束	说 明
receiver	VARCHAR	20		收货人
recvaddrid	INTEGER		FOREIGN KEY	收货地址编码
recvaddr	VARCHAR	128		收货地址
recvmobile	CHAR	11		收货人手机号码
shipper	VARCHAR	20		发货人
shipaddrid	INTEGER		FOREIGN KEY	发货地址编码
shipaddr	VARCHAR	128		发货地址
shipmobile	CHAR	11		发货人手机号码
trackno	INTEGER			快递单号
exprname	VARCHAR	128		公司名称
courname	VARCHAR	20		快递员姓名
courmobile	CHAR	11		手机号码
mot	CHAR	4		运输方式
doe	DATE			预计到达日期
doa	DATE			送达日期

表 1.8 Lineitems 表结构及其约束

字 段 名	数 据 类 型	长度	约 束	说 明
orderid	INTEGER		PRIMARY KEY/FOREIGN KEY	订单编号
shopid	INTEGER		PRIMARY KEY/FOREIGN KEY	店铺编号
goodid	INTEGER		PRIMARY KEY/FOREIGN KEY	商品编号
saleprice	MONEY			销售价格
saleamt	INTEGER			销售数量

表 1.9 Categories 表结构及其约束

字 段 名	数 据 类 型	长度	约 束	说 明
catgid	INTEGER		PRIMARY KEY	分类编码
stdcode	CHAR	14		国家标准编码
catgname	VARCHAR	128		分类名称
parentid	INTEGER			父类编码
currlevel	INTEGER			当前层级

表 1.10　Adminaddrs 表结构及其约束

字 段 名	数 据 类 型	长度	约　　束	说　　明
addrid	INTEGER		PRIMARY KEY	地址编码
stdcode	CHAR	14		国家标准编码
addrname	VARCHAR	40		地址名称
parentid	INTEGER			父地址编码
location	GEOMETRY			经度纬度合一起的 GIS 数据
currlevel	INTEGER			当前层级

1.4.2　创建表

（1）创建 Shopstores 表。

程序代码如下。

```
CREATE TABLE shopstores (
  shopid character(6 char) NOT NULL,
  shopname character varying(128 char) NULL,
  shopurl character varying(256 char) NULL,
  custgrading numeric(5,2) NULL,
  delygrading numeric(5,2) NULL,
  servgrading numeric(5,2) NULL,
  comprgrading numeric(5,2) NULL,
  CONSTRAINT shopstores_pkey PRIMARY KEY (shopid)
);
```

（2）创建 Goods 表。

程序代码如下。

```
CREATE TABLE goods (
  goodid character(12 char) NOT NULL,
  goodname character varying(128 char) NULL,
  mfrs character varying(256 char) NULL,
  brand character varying(20 char) NULL,
  model character varying(50 char) NULL,
  grossweight real NULL,
  poo character varying(128 char) NULL,
  catgid integer NULL,
  price money NULL,
  dop date NULL,
  doe date NULL,
  shelflife interval NULL,
  features json NULL,
  size polygon NULL,
  instmanual text NULL,
```

```
  CONSTRAINT goods_pkey PRIMARY KEY (goodid)
);
ALTER TABLE goods ADD CONSTRAINT goods_catgid_fkey FOREIGN KEY (catgid)
REFERENCES categories(catgid);
```

(3) 创建 Supply 表。

程序代码如下。

```
CREATE TABLE supply (
  shopid character(6 char) NOT NULL,
  goodid character(12 char) NOT NULL,
  totlwhamt integer NULL,
  price money NULL,
  discount numeric(5,2) NULL,
  status boolean NULL,
  url character varying(512 char) NULL,
  homepage xml NULL,
  CONSTRAINT supply_pkey PRIMARY KEY (shopid, goodid)
);
ALTER TABLE supply ADD CONSTRAINT supply_goodid_fkey FOREIGN KEY (goodid)
REFERENCES goods(goodid);
ALTER TABLE supply ADD CONSTRAINT supply_shopid_fkey FOREIGN KEY (shopid)
REFERENCES shopstores(shopid);
```

(4) 创建 Customers 表。

程序代码如下。

```
CREATE TABLE customers (
  custid integer NOT NULL,
  custname character varying(20 char) NULL,
  gender character(2 char) NULL,
  dob date NULL,
  email character varying(128 char) NULL,
  ipaddress character varying(128 char) NULL,
  mobile character(11 char) NULL,
  address character varying(128 char) NULL,
  mi int8range NULL,
  ebg character(4 char) NULL,
  prof character(50 char) NULL,
  CONSTRAINT customers_ebg_check CHECK ((ebg = ANY (ARRAY['小学'::bpchar, '初中'::bpchar, '高中'::bpchar, '中专'::bpchar, '大专'::bpchar, '本科'::bpchar, '硕士'::bpchar, '博士'::bpchar, '其他'::bpchar]))),
  CONSTRAINT customers_gender_check CHECK ((gender = ANY (ARRAY['男'::bpchar, '女'::bpchar]))),
  CONSTRAINT customers_pkey PRIMARY KEY (custid)
```

(5) 创建 Comments 表。

程序代码如下。

```
CREATE TABLE comments (
    orderid character(12 char) NOT NULL,
    goodid character(12 char) NOT NULL,
    custid integer NOT NULL,
    stars integer NULL,
    remark character varying(512 char) NULL,
    feeling character(4 char) NULL,
    submtime timestamp with time zone NULL,
    CONSTRAINT comments_check CHECK (((((feeling = '好评'::bpchar) AND (stars >=
4)) OR ((feeling = '中评'::bpchar) AND ((stars = 3) OR (stars = 2)))) OR ((feeling
= '差评'::bpchar) AND (stars = 1)))),
    CONSTRAINT comments_pkey PRIMARY KEY (orderid, goodid, custid)
);
ALTER TABLE comments ADD CONSTRAINT comments_orderid_fkey FOREIGN KEY (orderid)
REFERENCES orders(orderid);
ALTER TABLE comments ADD CONSTRAINT comments_custid_fkey FOREIGN KEY (custid)
REFERENCES customers(custid);
ALTER TABLE comments ADD CONSTRAINT comments_goodid_fkey FOREIGN KEY (goodid)
REFERENCES goods(goodid);
```

(6) 创建 Orders 表。

程序代码如下。

```
CREATE TABLE orders (
    ordid character(12 char) NOT NULL,
    custid integer NULL,
    shopid character(6 char) NULL,
    submtime timestamp with time zone NULL,
    ipaddr inet NULL,
    totlbal money NULL,
    carriage money NULL,
    discamt money NULL,
    finlbal money NULL,
    paytime timestamp with time zone NULL,
    paymethod character(10 char) NULL,
    receiver character varying(20 char) NULL,
    recvaddrid integer NULL,
    recvaddr character varying(128 char) NULL,
    recvmobile character(11 char) NULL,
    shipper character varying(20 char) NULL,
    shipaddrid integer NULL,
    shipaddr character varying(128 char) NULL,
    shipmobile character(11 char) NULL,
    trackno character(11 char) NULL,
    exprname character varying(128 char) NULL,
    courname character varying(20 char) NULL,
    courmobile character(11 char) NULL,
    mot character(4 char) NULL,
    doe date NULL,
```

```
    doa date NULL,
    CONSTRAINT orders_pkey PRIMARY KEY (ordid)
);

ALTER TABLE orders ADD CONSTRAINT orders_custid_fkey FOREIGN KEY (custid)
REFERENCES sales.customers(custid);
ALTER TABLE orders ADD CONSTRAINT orders_recvaddrid_fkey FOREIGN KEY
(recvaddrid) REFERENCES adminaddrs(addrid);
ALTER TABLE orders ADD CONSTRAINT orders_shipaddrid_fkey FOREIGN KEY
(shipaddrid) REFERENCES adminaddrs(addrid);
ALTER TABLE orders ADD CONSTRAINT orders_shopid_fkey FOREIGN KEY (shopid)
REFERENCES sales.shopstores(shopid);
```

（7）创建 Lineitems 表。

程序代码如下。

```
CREATE TABLE lineitems (
    ordid character(12 char) NOT NULL,
    shopid character(6 char) NOT NULL,
    goodid character(14 char) NOT NULL,
    saleprice money NULL,
    saleamt integer NULL,
    CONSTRAINT lineitems_pkey PRIMARY KEY (ordid, shopid, goodid)
);

ALTER TABLE lineitems ADD CONSTRAINT lineitems_goodid_fkey FOREIGN KEY (goodid)
REFERENCES goods(goodid);
ALTER TABLE lineitems ADD CONSTRAINT lineitems_ordid_fkey FOREIGN KEY (ordid)
REFERENCES orders(ordid);
ALTER TABLE lineitems ADD CONSTRAINT lineitems_shopid_fkey FOREIGN KEY (shopid)
REFERENCES shopstores(shopid);
ALTER TABLE lineitems ADD CONSTRAINT lineitems_shopid_goodid_fkey FOREIGN KEY
(shopid, goodid) REFERENCES supply(shopid, goodid);
```

（8）创建 Categories 表。

程序代码如下。

```
CREATE TABLE categories (
    catgid integer NOT NULL,
    stdcode character(14 char) NULL,
    catgname character varying(128 char) NULL,
    parentid integer NULL,
    currlevel integer NULL,
    CONSTRAINT categories_pkey PRIMARY KEY (catgid),
    CONSTRAINT categories_parentid_fkey FOREIGN KEY (parentid) REFERENCES
categories(catgid)
);
```

（9）创建 Adminaddrs 表。

程序代码如下。

```
CREATE TABLE adminaddrs (
  addrid character varying(12 char) NOT NULL,
  stdcode character(14 char) NULL,
  addrname character varying(40 char) NULL,
  parentid integer NULL,
  currlevel integer NULL,
  CONSTRAINT adminaddrs_pkey PRIMARY KEY (addrid),
  CONSTRAINT adminaddrs_parentid_fkey FOREIGN KEY (parentid) REFERENCES adminaddrs(addrid)
);
```

建好表以后，可以向表中单行或批量插入数据。为了后续章节设计实例所用，本书对上述各表分别插入了相关测试数据，具体操作不再详述。

此外，本书中所有示例，如果不加特殊说明，均默认在 sales 模式下运行。

1.5 使用金仓在线帮助文档

PL/SQL 的最新技术通常都通过金仓官方文档发布，学会查阅、使用金仓官方文档，是学好 PL/SQL 的有效途径。人大金仓为开发人员和 DBA 提供了大量详实的文档，以辅助参考、学习。这些官方文档都发布在金仓官网上，用户注册、登录后就可以下载。网址为：https://help.kingbase.com.cn。

其中，与 PL/SQL 相关的文档包括如下内容。

（1）KingbaseES SQL 语言参考手册。

（2）KingbaseES PL/SQL 过程语言参考手册。

（3）KingbaseES 数据库开发指南。

（4）KingbaseES 数据库概念。

（5）KingbaseES 数据库性能调优指南。

第 2 章

PL/SQL 程序结构

PL/SQL 程序结构由基本的语句块或语句块之间相互嵌套组合而成。语句块中包含的控制语句是其基本组成元素,用以控制程序的运行逻辑。PL/SQL 有 3 类控制语句:条件选择语句、循环语句和顺序控制语句。其中,条件选择语句根据不同的条件运行不同的逻辑处理;循环语句根据不同的循环条件执行相同的循环体;顺序控制语句通过 GOTO 语句进行程序流程的转移。本章将主要介绍 PL/SQL 的程序结构以及 3 种主要控制语句。

- PL/SQL 块结构。
- PL/SQL 声明段。
- 基本执行语句。
- 条件选择语句。
- 循环语句。
- 异常捕获。
- 获取执行状态信息。

2.1 PL/SQL 块结构

在 PL/SQL 程序中,和其他大多数结构化程序语言一样,基本的代码单元是语句块,所有的 PL/SQL 程序都是由语句块构成的。

一个完整的 PL/SQL 语句块由 3 部分组成。代码如下。

```
[ <<label>> ]                    --语句块标识(可选)
[DECLARE                         --声明部分(可选)
    declarations ]
BEGIN                            --执行部分(必需)
    statements                   --语句(可以使用声明部分的元素)
[EXCEPTION                       --异常处理部分(可选)
    statements ]                 --异常处理操作(处理执行部分抛出的异常)
END [ label ];
```

其中,

(1) 声明部分:声明部分是可选的,以关键字 DECLARE 开始,以 BEGIN 结束。该部分主要用于声明变量、常量、数据类型、游标、异常处理名称和本地(局部)子程序定义等。变量只有经过声明后才能在执行部分使用。

(2) 执行部分：是 PL/SQL 块的功能实现部分，以关键字 BEGIN 开始，以 END 结束。该部分通过变量赋值、数据查询、数据操纵、数据定义、事务控制、游标处理等操作实现块的功能。也可以使用各种控制结构，嵌入各种 SQL 语句，实现更复杂的业务逻辑功能。

(3) 异常处理部分：以关键字 EXCEPTION 开始，以 END 结束。该部分用于捕获执行部分产生的异常，如果没有产生异常或者异常没有被捕获，则该部分都不会执行。

(4) label 可以用来标识一个块，以便在一个 EXIT 语句中使用或者标识在该块中声明的变量名。位于 END 之后的 label 必须与块开始的 label 相匹配。

定义 PL/SQL 块时需要注意下列事项：

(1) 执行部分是必需的，而声明部分、异常部分是可选的。

(2) 在 PL/SQL 语句块中的每个声明和语句都以一个分号结束。

(3) 可以在一个块的声明部分嵌套命名块，即嵌套函数。

(4) 可以在一个块的执行部分或异常处理部分嵌套其他的 PL/SQL 块。

(5) 所有的 PL/SQL 块都以"END;"结束。

(6) PL/SQL 中通过标识符命名变量、常量、游标、异常、关键字、标签、包、保留字、子程序、数据类型等，标识符不区分大小写。同时，命名时要注意与保留字、关键字的不同。

(7) PL/SQL 中支持对运算符的广泛使用，如+、-、=、>=等，具体内容请参考金仓在线帮助文档。

(8) PL/SQL 支持使用注释，分为单行注释和多行注释两种。其中，单行注释可以在一行的任何地方以"--"开始，直到该行结尾；多行注释以"/*"开始，以"*/"结束。

示例 2.1：一个完整的 PL/SQL 语句块。

功能描述：从 adminaddrs 表中查找大连市对应的行政区域 id。

程序代码如下。

```
DECLARE   SAL_TABLE   VARCHAR2;
BEGIN
    SELECT   ADDRID   INTO   SAL_TABLE   FROM ADMINADDRS
    WHERE ADDRNAME='大连市';
    RAISE NOTICE 'SAL_TABLE=%',SAL_TABLE ;
EXCEPTION
    WHEN NO_DATA_FOUND
        THEN RAISE NOTICE '%','没有查到相关数据。';
END;
```

基本的代码单元是语句块，语句块是可以嵌套的，即嵌套另一个 PL/SQL 语句块，称为子块。子块可以用来进行逻辑分组或将变量局部化为语句的一个小的语句组。在子块的生命期内，其所声明的变量会掩盖外层块中相同名称的变量，即存在变量的作用域和可见性问题。

1. 变量的作用域

(1) 变量的作用域是指变量的有效作用范围从变量声明开始，直到块结束。

(2) 如果 PL/SQL 块相互嵌套，则在内部块中声明的变量是局部的，只能在内部块中引用;在外部块中声明的变量是全局的,既可以在外部块中引用,也可以在内部块中引用。

(3) 如果内部块与外部块中定义了同名变量,则在内部块中引用外部块的全局变量,需要使用外部块名进行标识。

2. 变量的可见性

(1) 变量的可见性是指在 PL/SQL 单元中可以直接引用的变量。

(2) 如果 PL/SQL 单元相互嵌套,并且在内部块中声明了与外部块中声明的变量同名,则在内部块中,同名的外部变量(全局变量)与内部变量(局部变量)都有效,但外部变量不可见。如要引用外部变量,应给外部块定义一个标签,然后通过外部块的标签指定要引用的外部变量。如果外部块没有标签,则在内部块中不能引用该同名全局变量。

需要注意的是,在一个 PL/SQL 块中不能引用其他同级的 PL/SQL 块中声明的变量。

示例 2.2:变量作用域与可见性示例。

程序代码如下。

```
--外部块
DECLARE
    A CHAR;                         --A(CHAR)的作用域开始
    B NUMBER;                       --B 的作用域开始
BEGIN
    /* A(CHAR), B 可见 */
    --第 1 个子块
    DECLARE
        A INTEGER;                  --A(INTEGER)作用域开始
        C NUMBER;                   --C 的作用域开始
    BEGIN
        NULL;                       --A(INTEGER), B, C 可见
    END;                            --A(INTEGER)和 C 的作用域结束

    --第 2 个子块
    DECLARE
        D NUMBER;                   --D 的作用域开始
    BEGIN
        NULL;                       --A(CHAR), B, D 可见
    END;                            --D 的作用域结束
                                    --A(CHAR), B 可见
END;
```

示例 2.3:内部块引用外部块全局变量示例。

程序代码如下。

```
BEGIN
  <<OUTER_BLOCK>>
  DECLARE
    V_CITY VARCHAR2(20);            --定义外层块变量
  BEGIN
    V_CITY:='沈阳市';               --为外层块的变量赋初值
```

```
   <<INNER_BLOCK>>
   DECLARE
      V_CITY VARCHAR2(20); --定义与外层块同名的内层块变量
   BEGIN
      V_CITY:='大连市';                              --为内层块变量赋值
      /*输出内层块的变量*/
      RAISE NOTICE '内层块的城市名称:%',V_CITY;
      /*在内层块中访问外层块的变量*/
      RAISE NOTICE '外内层块的城市名称:%',OUTER_BLOCK.V_CITY;
   END INNER_BLOCK;
   RAISE NOTICE 'OUTER城市名称:%',V_CITY;           --在外层块中访问变量
 END OUTER_BLOCK;
END;
```

2.2　PL/SQL 声明段

在声明段部分，为值分配存储空间，指定其数据类型，并命名存储位置，以便后续可以引用它。在 PL/SQL 中，必须先声明对象，然后才能引用它们。声明段可以出现在任何块、子程序或包的声明部分。

另外，在声明段中声明的变量、常量、参数、函数的返回值时，都需要指定具体的数据类型，以确定数据的存储格式、有效取值以及可以进行操作的类型。数据类型具有特定的属性，对不同数据类型的值，KingbaseES 的处理也不同。

2.2.1　数据类型

每个 PL/SQL 常量、变量、参数和函数返回值都有一个数据类型，决定了它的存储格式以及它的有效值和操作。PL/SQL 预定义了一套完善的标量和复合数据类型，也可以创建用户自定义类型。

（1）标量类型：用于存储单个值、内部没有子组件的数据类型。PL/SQL 标量数据类型可以进一步划分为以下几种。

① SQL 数据类型，如字符数据类型、数值类型、日期类型、布尔类型等。

② REF CURSOR 类型。

③ PLS_INTEGER 等 PL/SQL 预定义的子类型。

④ 用户定义的 PL/SQL 子类型。

（2）复合类型：复合类型是指内部包含子组件的数据类型，每个子组件都是一个标量或复合类型。如记录类型、联合数组类型、嵌套表类型、可变数组类型等。

子类型是在标准类型上添加一些限制生成的新类型，是其基本类型的子集。PLS_INTERGER 就是 PL/SQL 预定义的子类型，PL/SQL 还允许用户定义自己的子类型，语法格式如下。

```
SUBTYPE subtype_name IS type_name[(constraint)] [NOT NULL]
```

subtype_name 指定子类型的名称。

type_name 指定子类型所基于的原始类型的名称。原始类型可以是任何标量或用户定义的 PL/SQL 数据类型说明符,如 CHAR、DATE 或 RECORD(包括先前定义的用户定义子类型)。

constraint 子句定义支持精度或小数位数的类型的限制。语义描述如下。

```
{precision [, scale]} | length
```

precision 指定子类型值中允许的总位数。
scale 指定子类型值中允许的小数位数。
length 指定 CHARACTER、VARCHAR 或 TEXT 基本类型值中允许的总长度。

2.2.2 变量和常量的声明

如果需要在 PL/SQL 程序中使用变量或常量,必须先在 PL/SQL 块的声明部分声明该变量或常量。语法格式如下。

```
variable_name [CONSTANT] datatype [NOT NULL] [DEFAULT|:=expression];
```

说明如下。
(1) 变量或常量名称是一个 PL/SQL 标识符,应符合标识符命名规范。
(2) 每次声明只能定义一个变量或者常量。
(3) 如果加上关键字 CONSTANT,则表示所定义的是一个常量,必须为它赋初值。
(4) 如果定义变量时使用了 NOT NULL 关键字约束,则必须为变量赋初值。
(5) 如果变量没有赋初值,则默认值为 NULL。
(6) 使用"="或":="运算符为变量初始化。

示例 2.4:带有初始值的变量和常量声明。
程序代码如下。

```
DECLARE
  v1 NUMBER(4);
  v2 NUMBER(4) NOT NULL :=10;
  v3 CONSTANT NUMBER(4) DEFAULT 100;
BEGIN
  IF v1 IS NULL THEN
    raise  notice 'V1 IS NULL! ';
  END IF;
  raise notice 'v2=%,v3=%',v2,v3;
END;
```

程序运行结果如下。

```
NOTICE: V1 IS NULL!
NOTICE:v2=10,v3=100
```

声明部分除了直接指定确定的数据类型外,还可以使用%TYPE 属性声明。%TYPE 属性可以声明与之前声明的变量或列具有相同数据类型的数据项(不需要知道该类型具体是什么)。如果被引用项的声明被更改,则引用项的声明也会相应更改。该声明的语法格式如下。

```
referencing_item referenced_item%TYPE;
```

引用项从被引用项继承数据类型和大小,被引用项的约束条件不被继承。当声明变量用来抓取数据库数值时,％TYPE 属性特别有用。声明与列类型相同的变量,语法格式如下。

```
variable_name table_name.column_name%TYPE;
```

示例 2.5:使用％TYPE 属性声明变量。
程序代码如下。

```
DECLARE
  v1 categories.catgid%TYPE;
  v2 categories.catgname%TYPE:='手机';
  v3 v2%TYPE;
BEGIN
  IF v1 IS NULL THEN
    raise  notice 'V1 IS NULL! ';
  END IF;
  raise notice 'v2=%,v3=%',v2,v3;
END;
```

程序运行结果如下。

```
NOTICE:V1 IS NULL!
NOTICE:v2=手机,v3=<NULL>
```

2.3 基本执行语句

PL/SQL 的基本执行语句包括赋值语句、查询语句、DML 语句、RETURN 语句、GOTO 语句、NULL 语句等。

1. 赋值语句

声明变量后,可以通过以下两种方式为其赋值。
(1) 使用赋值语句将表达式赋值给变量。
(2) 使用 SELECT INTO 或 FETCH 语句从 table 中赋值给变量。
将表达式的值赋给变量,语法声明如下。

```
variable_name := expression;
```

SELECT INTO 语句的一种简单形式如下。

```
SELECT select_item [, select_item ]...
INTO variable_name [, variable_name ]...
FROM table_name;
```

需要注意的是,变量和值必须具有兼容的数据类型。如果一种数据类型可以隐式转换为另一种数据类型,那么它与另一种数据类型必须兼容。

2. 查询语句

查询语句(SELECT)是从一个或更多表中返回记录行。与标准的 SELECT 语句不同,在 PL/SQL 程序中,SELECT 语句需要与 INTO 或 BULK COLLECT INTO 短语结合使用,将查询结果保存到变量中。针对单条和多条记录的查询有以下两种方式。

(1) SELECT INTO:单条记录的查询。

(2) SELECT BULK COLLECT INTO:多条记录的查询。

INTO 句子后的变量用于接收查询的结果,变量的个数、顺序应该与查询的目标数据相匹配,也可以是记录类型的变量。

3. DML 语句

PL/SQL 中的 DML 语句对标准 SQL 语句中的 DML 语句(如 INSERT、UPDATE、DELETE、MERGE)进行扩展,允许使用变量。有关语法请参阅金仓在线帮助文档。

4. RETURN 语句

RETURN 语句用于立即结束子程序或包含它的匿名块的执行。

在函数中,RETURN 语句将指定的值分配给函数标识符,并将控制权返回给调用者,调用者调用后立即恢复执行(可能在调用语句内)。函数中的每个执行路径都必须指向一个 RETURN 语句,否则 PL/SQL 运行时会报错。

在过程中,RETURN 语句将控制权返回给调用者,调用后立即恢复执行。在匿名块中,RETURN 语句退出它自己的块和所有封闭块。

一个子程序或匿名块可以包含多个 RETURN 语句。

5. GOTO 语句

GOTO 语句无条件地将控制跳转到标签。标签在其范围内必须是唯一的,并且必须位于可执行语句或 PL/SQL 块之前。运行时,GOTO 语句将控制跳转到带标签的语句或块。

需要谨慎使用 GOTO 语句,过度使用它们将导致代码难以理解和维护。不要使用 GOTO 语句将控制从深度嵌套的结构跳转到异常处理程序,这将引发异常。有关 PL/SQL 异常处理机制的信息,请参阅 PL/SQL 异常处理。

对 GOTO 声明的限制如下。

(1) 如果 GOTO 语句过早退出 FOR LOOP 语句,游标将关闭。

(2) GOTO 语句不能将控制转移到 IF 语句、CASE 语句、LOOP 语句或子块中。

(3) GOTO 语句不能将控制从一个 IF 语句子句转移到另一个语句,或者从一个 CASE 语句转移到另一个语句。

(4) GOTO 语句不能将控制转移出子程序。

(5) GOTO 语句不能将控制转移到异常处理程序中。

（6）GOTO 语句不能将控制从异常处理程序转移到当前块中(但它可以将控制从异常处理程序转移到封闭块中)。

语法格式如下。

```
goto_statement ::=
GOTO label ;
```

6. NULL 语句

NULL 语句只将控制权传递给下一条语句,一些语言将这样的指令称为无操作。

NULL 语句的用途包括如下几项。

（1）为 GOTO 语句提供目标。
（2）通过明确条件语句的含义和动作来提高可读性。
（3）创建占位符和存根子程序。
（4）表明开发人员意识到某种可能性,但无须采取任何行动。

示例 2.6：综合性示例。

功能描述：根据 id 判断记录是否存在,如果存在,则添加,否则更新相关记录;删除记录时应该先判断是否存在。

程序代码如下。

```
DECLARE
    shop_id shopstores.shopid %TYPE := 2;
    shop_name shopstores.shopname %TYPE := 'xm';
    shop_comprgrading shopstores.comprgrading %TYPE := 8.00;
    shop_name1 shopstores.shopname %TYPE;
BEGIN
    IF NOT EXISTS(SELECT shopid FROM sales.shopstores WHERE shopid = shop_id) THEN
        RAISE NOTICE '插入一条记录';
        GOTO INSERT;
    ELSE
        RAISE NOTICE '更新该记录';
        GOTO UPDATE;
    END IF;

    <<INSERT>>
        INSERT INTO sales.shopstores
        VALUES (shop_id , shop_name, NULL, NULL, NULL, NULL, shop_comprgrading);

    <<UPDATE>>
        UPDATE shopstores
        SET shopname =    shop_name, comprgrading = shop_comprgrading
        WHERE shopid = shop_id ;

    SELECT shopname INTO shop_name1
    FROM "sales"."shopstores"
    WHERE shopid = shop_id;
```

```
    RAISE NOTICE 'shop_name1:%',shop_name1;

    IF EXISTS (SELECT shopid FROM shopstores WHERE shopid = 2) THEN
      DELETE FROM sales.shopstores
      WHERE shopid = 2;
      RETURN;
      RAISE NOTICE '删除成功';
    ELSE
      RAISE NOTICE '该货物不存在';
      NULL;
    END IF;
END;
```

程序运行结果如下。

```
插入一条记录
shop_name1:xm
```

2.4 条件选择语句

PL/SQL 中,条件选择语句包括 IF 和 CASE,二者针对不同的数据值,运行形式也不同。

(1) IF 语句根据条件运行或跳过一个或多个语句。IF 语句有以下形式。

① IF THEN。

② IF THEN ELSE。

③ IF THEN ELSIF。

(2) CASE 语句从一系列条件中进行选择,并运行相应的语句。CASE 语句有以下形式。

① 简单 CASE 语句,它计算单个表达式,并将其与几个值进行比较。

② 搜索 CASE 语句,它计算多个条件,并选择第一个为真的条件。

当存在多个备选方案,并需要采取不同操作时,CASE 语句更适合。

顺序控制语句包括 GOTO 语句和 NULL 语句。GOTO 语句实现程序执行流程的跳转,但容易破坏程序执行的过程化,不利于程序的维护和可读性,一般不建议使用。NULL 语句不作任何操作,起到占位符作用,除了提高程序可读性外,也为程序的扩展作了适当考虑。

2.4.1 IF

IF 语句运行或跳过一个或多个语句的序列,具体取决于 BOOLEAN 表达式的值。分为下列 3 种情况。

(1) IF THEN。

(2) IF THEN ELSE。

(3) IF THEN ELSIF。

语法格式如下。

```
if_statement ::=
IF boolean_expression THEN statement [ statement ]...
[ ELSIF boolean_expression THEN statement [ statement ]... ]...
[ ELSE statement [ statement ]... ]
END IF ;
```

语义描述如下。

boolean_expression

① 值为 TRUE、FALSE 或 NULL 的表达式。

② 第一个 boolean_expression 总是被评估。仅当前面表达式的值为 FALSE 时，才会评估其他每个 boolean_expression。

③ 如果计算一个 boolean_expression 并且其值为 TRUE，则相应 THEN 之后的语句将运行。随后的表达式不会被计算，与它们关联的语句也不会运行。

ELSE

如果布尔表达式的值不是 TRUE，则运行 ELSE 之后的语句。

1．IF-THEN

IF THEN 语句根据条件运行或跳过一个或多个语句。

IF THEN 语句具有以下结构。

```
IF condition THEN
    statements
END IF;
```

如果 condition 为真，则 statements 运行；否则，statements 不运行。

示例 2.7：IF THEN 语句。

功能描述：在本例中，当且仅当 a 大于 b 时，打印 a−b 的值。

程序代码如下。

```
DECLARE
PROCEDURE p(a int, b int) AS
BEGIN
  IF a > b THEN
    RAISE NOTICE 'a - b = %', a - b;
  END IF;
END;

BEGIN
  p(1,2);
  p(3,2);
END;
```

程序运行结果如下。

```
NOTICE:  a - b = 1
```

2. IF-THEN-ELSE

IF THEN ELSE 语句具有以下结构。

```
IF condition THEN
   statements
ELSE
   else_statements
END IF;
```

如果条件的值为真，则 statements 运行；否则，else_statements 运行。

IF 语句可以嵌套，如示例 2.8 所示。

示例 2.8：IF THEN ELSE 语句。

功能描述：在本例中，当 a 大于 b 时，打印 a-b 的值，当 a 等于 b 时，打印"a=b"，当 a 小于 b 时，打印 a+b 的值。

程序代码如下。

```
DECLARE
PROCEDURE p(a int, b int) AS
BEGIN
   IF a > b THEN
       RAISE NOTICE 'a - b = %', a - b;
   ELSE
       IF a = b THEN
           RAISE NOTICE 'a = b';
       ELSE
           RAISE NOTICE 'a + b = %', a + b;
       END IF;
   END IF;
END;

BEGIN
    p(1, 2);
    p(2, 2);
    p(3, 2);
END;
```

程序运行结果如下。

```
NOTICE: a + b = 3
NOTICE: a = b
NOTICE: a - b = 1
```

3. IF-THEN-ELSIF

IF THEN ELSIF 语句具有以下结构。

```
IF condition_1 THEN
   statements_1
ELSIF condition_2 THEN
   statements_2
[ ELSIF condition_3 THEN
   statements_3
]...
[ ELSE
   else_statements
]
END IF;
```

IF THEN ELSIF 语句运行条件为真的第一条语句,不计算剩余条件。如果没有条件为真,则运行 else_statements(如果存在);否则,IF THEN ELSIF 不执行任何语句。

相比等效嵌套的 IF THEN ELSE 语句,单个 IF THEN ELSIF 语句更易于理解。代码如下。

```
-- IF THEN ELSIF statement
IF condition_1 THEN statements_1;
   ELSIF condition_2 THEN statements_2;
   ELSIF condition_3 THEN statement_3;
END IF;

-- Logically equivalent nested IF THEN ELSE statements
IF condition_1 THEN
   statements_1;
ELSE
   IF condition_2 THEN
       statements_2;
   ELSE
       IF condition_3 THEN
       statements_3;
       END IF;
   END IF;
END IF;
```

示例 2.9:IF THEN ELSIF 语句。

功能描述:在本例中,使用 ELSIF 具有嵌套 IF THEN ELSE 同样的功能。

程序代码如下。

```
DECLARE
PROCEDURE p(a int, b int) AS
BEGIN
   IF a > b THEN
        RAISE NOTICE 'a - b = %', a - b;
   ELSIF a = b THEN
      RAISE NOTICE 'a = b';
   ELSE
```

```
      RAISE NOTICE 'a + b = %', a + b;
  END IF;
END;

BEGIN
  p(1, 2);
  p(2, 2);
  p(3, 2);
END;
```

程序运行结果如下。

```
NOTICE:  a + b = 3
NOTICE:  a = b
NOTICE:  a - b = 1
```

示例 2.10：IF THEN ELSIF 语句模拟简单的 CASE 语句。

功能描述：本例中使用带有许多 ELSIF 子句的 IF THEN ELSIF 语句,将单个值与许多可能值进行比较。相比而言,一个简单的 CASE 语句更清晰,见示例 2.11。

程序代码如下。

```
DECLARE
  grade CHAR(1);
BEGIN
  grade := 'A';

  IF grade = 'A' THEN
      RAISE NOTICE 'Excellent';
  ELSIF grade = 'B' THEN
      RAISE NOTICE 'Very Good';
  ELSIF grade = 'C' THEN
      RAISE NOTICE 'Good';
  ELSIF grade = 'D' THEN
      RAISE NOTICE 'Fair';
  ELSIF grade = 'F' THEN
      RAISE NOTICE 'Poor';
  ELSE
      RAISE NOTICE 'No such grade';
  END IF;
END;
```

程序运行结果如下。

```
NOTICE: Excellent
```

2.4.2 CASE

CASE 语句从一系列条件中进行选择,并运行相应的语句。

简单 CASE 语句计算单个表达式,并将其与几个潜在值进行比较。

搜索 CASE 语句计算多个布尔表达式,并选择第一个值为 TRUE 的表达式。

1. 简单 CASE 语句

简单 CASE 语句具有以下结构。

```
CASE selector
WHEN selector_value_1 THEN statements_1
WHEN selector_value_2 THEN statements_2
...
WHEN selector_value_n THEN statements_n
[ ELSE
   else_statements ]
END CASE;]
```

selector 是一个表达式(通常是单个变量)。每个 selector_value 可以是文字或表达式(对于完整的语法,请参阅"CASE")。

如果 selector 等于第一条语句值 selector_value_1,则不计算剩余条件。如果没有任何 selector_value 等于 selector,则执行 else_statements 语句,如果没有该语句,则引发预定义的异常 CASE_NOT_FOUND。

示例 2.11 使用一个简单的 CASE 语句将单个值与多个可能值进行比较。请读者与示例 2.10 中的 IF THEN ELSIF 语句进行对比。

注意:如果简单 CASE 语句中的 selector 的值为 NULL,则它不能被 WHEN NULL 匹配,而需要使用带有 WHEN 条件 IS NULL 的搜索 CASE 语句。

示例 2.11:简单的 CASE 语句。

程序代码如下。

```
DECLARE
   grade CHAR(1);
BEGIN
   grade := 'B';

   CASE grade
       WHEN 'A' THEN RAISE NOTICE 'Excellent';
       WHEN 'B' THEN RAISE NOTICE 'Very Good';
       WHEN 'C' THEN RAISE NOTICE 'Good';
       WHEN 'D' THEN RAISE NOTICE 'Fair';
       WHEN 'F' THEN RAISE NOTICE 'Poor';
       ELSE RAISE NOTICE 'No such grade';
   END CASE;
END;
```

程序运行结果如下。

```
NOTICE:  Very Good
```

2. 搜索 CASE 语句

搜索 CASE 语句具有以下结构。

```
CASE
WHEN condition_1 THEN statements_1
WHEN condition_2 THEN statements_2
...
WHEN condition_n THEN statements_n
[ ELSE
  else_statements ]
END CASE;]
```

搜索 CASE 语句运行条件为真的第一条语句，不计算剩余条件。如果没有条件为真，即 CASE 语句如果存在 else_statements，则运行，否则引发预定义异常 CASE_NOT_FOUND。

示例 2.12 中的搜索 CASE 语句在逻辑上等价于例 2.11 中的简单 CASE 语句。

示例 2.12：搜索 CASE 语句。

程序代码如下。

```
DECLARE
  grade CHAR(1);
BEGIN
  grade := 'B';

  CASE
     WHEN grade = 'A' THEN RAISE NOTICE 'Excellent';
     WHEN grade = 'B' THEN RAISE NOTICE 'Very Good';
     WHEN grade = 'C' THEN RAISE NOTICE 'Good';
     WHEN grade = 'D' THEN RAISE NOTICE 'Fair';
     WHEN grade = 'F' THEN RAISE NOTICE 'Poor';
     ELSE RAISE NOTICE 'No such grade';
  END CASE;
END;
```

程序运行结果如下。

```
NOTICE: Very Good
```

2.5 循环语句

循环语句通过使用一系列不同的值迭代运行相同的语句。一个循环语句包含以下 3 部分：

（1）迭代变量。

（2）迭代器。

(3) 循环执行体。

语法结构如下。

```
loop_statement ::= [ iteration_scheme ]
LOOP
    loop_body
END LOOP [ label ];
iteration_scheme ::= WHILE expression | FOR iterator
```

循环语句包括以下内容。

(1) 基本的 LOOP。

(2) FOR LOOP。

(3) WHILE LOOP。

(4) FOREACH。

退出循环的语句包括以下内容。

(1) EXIT。

(2) EXIT WHEN。

退出当前循环迭代的语句包括以下内容。

(1) CONTINUE。

(2) CONTINUE WHEN。

EXIT、EXIT WHEN、CONTINUE 和 CONTINUE WHEN 可以出现在循环内的任何位置，但不能出现在循环体外。建议使用这些语句而不是 GOTO 语句，它们可以通过将控制转移到循环外的语句实现循环退出，或结束循环的当前迭代。

运行过程中引发的异常也会退出循环。

LOOP 语句可以被标记或嵌套。建议为嵌套循环使用标签，以提高可读性。此外，必须确保 END LOOP 语句中的标签与同一循环语句开头的标签匹配（编译器不作检查）。

2.5.1 基本循环语句

使用 LOOP、EXIT、CONTINUE、WHILE、FOR 和 FOREACH 语句，可以使 PL/SQL 函数重复执行一系列命令。

(1) LOOP。

语法格式如下。

```
[ label ] LOOP
    statements
END LOOP [ label ];
```

说明如下。

① LOOP 定义一个无条件的循环，它会无限重复，直到被 EXIT 或 RETURN 语句终止。

② 可选的 label 可以被 EXIT 和 CONTINUE 语句用在嵌套循环中，指定这些语句引用的是哪一层循环。

(2) EXIT。

语法格式如下。

```
EXIT [ label ] [ WHEN boolean-expression ];
```

说明如下。

① 如果没有给出 label，那么最内层的循环会被终止，然后 END LOOP 后面的语句会被执行。如果给出了 label，那么它必须是当前或更高层的嵌套循环或语句块的标签。然后该命名循环或块就会被终止，并且控制会转移到该循环/块相应的 END 之后的语句上。

② 如果指定了 WHEN，只有 boolean-expression 为 true 时才会发生循环退出。否则，控制会转移到 EXIT 之后的语句。

③ EXIT 可以被用在所有类型的循环中，它并不限于在无条件循环中使用。

④ 在和 BEGIN 块一起使用时，EXIT 会把控制权交给块结束后的下一个语句。

(3) CONTINUE。

语法格式如下。

```
CONTINUE [ label ] [ WHEN boolean-expression ];
```

说明如下。

① 如果没有给出 label，最内层循环的下一次迭代会开始，即循环体中剩余的所有语句将被跳过，并且控制返回到循环控制表达式（如果有），以决定是否需要另一次循环迭代。如果 label 存在，它指定应该继续执行的循环标签。

② 如果指定了 WHEN，该循环的下一次迭代只有在 boolean-expression 为真时才会开始。否则，控制会传递给 CONTINUE 后面的语句。

③ CONTINUE 可以被用在所有类型的循环中，它并不限于在无条件循环中使用。

(4) WHILE。

语法格式如下。

```
[ <<label>> ]
WHILE boolean-expression LOOP
    statements
END LOOP [ label ];
```

说明如下。

只要 boolean-expression 被计算为 true，WHILE 语句就会重复一个语句序列。每次进入到循环体之前都会检查该表达式。

(5) FOR。

语法格式如下。

```
[ <<label>> ]
FOR name IN [ REVERSE ] expression .. expression [ BY expression ] LOOP
    statements
END LOOP [ label ];
```

说明如下。

① FOR 循环在给定整数范围内迭代。

② 变量 name 自动定义为类型 integer，并且只在循环内存在（任何该变量名的现有定义在此循环内都将被忽略）。

③ 给出范围上下界的两个表达式在进入循环的时候计算一次。

④ 如果下界大于上界（或在 REVERSE 情况下相反），循环体根本不会被执行。而且不会抛出任何错误。

⑤ 如果没有指定 BY 子句，迭代步长为 1，否则步长是 BY 中指定的值，该值也只在循环进入时计算一次。

⑥ 如果指定了 REVERSE，那么每次迭代后步长值会被减除，而不是增加。

⑦ 如果一个 label 被附加到 FOR 循环，那么整数循环变量可以用 label 的限定名引用。

示例 2.13：使用 while loop、continue 语句和 exit 语句输出 11~19，跳过 15，当等于 19 时跳出循环。

程序代码如下。

```
DECLARE
  v_num number(10);
BEGIN
  v_num := 10;
  WHILE v_num < 20 LOOP
    EXIT WHEN v_num = 19;
    v_num = v_num + 1;
    IF v_num = 15 THEN
      CONTINUE ;
    END IF ;
    RAISE NOTICE '%', v_num;
  END LOOP ;
END
```

程序运行结果如下。

```
11
12
13
14
16
17
18
19
```

2.5.2　FOR LOOP 语句

FOR LOOP 语句为循环索引的每个值运行一个或多个语句。

FOR LOOP 头指定迭代器，迭代器指定一个迭代和迭代控件。迭代控件向迭代对象提供一系列值，以便在循环体中使用。循环体中有语句执行迭代的每一个值。

可用的迭代控制包括如下内容。

① Stepped Range：一种迭代控件，可生成一系列步进数值。当未指定步长时，计数控制是步长为 1 的 integer 类型的步长范围。

② 单个表达式：计算单个表达式的迭代控件。

③ 重复表达式：重复计算单个表达式的迭代控件。

④ 游标：一个迭代控件，它从游标、游标变量或动态 SQL 生成所有记录。

FOR LOOP 语句的索引或迭代数被隐式或显式声明为循环体本地的变量。

循环中的语句可以读取迭代的值，但不能更改它。循环体外的语句不能引用迭代。在 FOR LOOP 语句运行后，迭代数未定义。循环迭代数有时称为循环计数器。

示例 2.14：与变量同名的 FOR LOOP 语句索引。

如果 FOR LOOP 语句的索引与匿名块中声明的变量同名，则本地隐式声明会隐藏另一个声明，如本例所示，程序代码如下。

```
DECLARE
  i NUMBER := 10;
BEGIN
  FOR i IN 1..5 LOOP
    RAISE NOTICE 'Inside loop, i = %', i;
  END LOOP;

  RAISE NOTICE 'Outside loop, i = %', i;
END;
```

程序运行结果如下。

```
NOTICE:  Inside loop, i = 1
NOTICE:  Inside loop, i = 2
NOTICE:  Inside loop, i = 3
NOTICE:  Inside loop, i = 4
NOTICE:  Inside loop, i = 5
NOTICE:  Outside loop, i = 10
```

示例 2.15：具有相同索引名称的嵌套 FOR LOOP 语句。

功能描述：在本例中，嵌套 FOR LOOP 语句的索引具有相同的名称。内循环通过使用外循环的标签引用外循环的索引。为清楚起见，内部循环还可以使用自身标签限定对本身索引的引用。

程序代码如下。

```
BEGIN
  <<label1>>
  FOR i IN 1..3 LOOP
    <<label2>>
    FOR i IN 1..3 LOOP
      IF label1.i = 2 THEN
        RAISE NOTICE 'lable1: i = %, label2: i = %',label1.i, label2.i;
```

```
        END IF;
      END LOOP label2;
  END LOOP label1;
END;
```

程序运行结果如下。

```
NOTICE:   lable1: i = 2, label2: i = 1
NOTICE:   lable1: i = 2, label2: i = 2
NOTICE:   lable1: i = 2, label2: i = 3
```

2.5.3 WHILE LOOP 语句

WHILE LOOP 语句在条件为 TRUE 时运行一个或多个语句。

WHILE LOOP 语句在条件变为 FALSE 或 NULL 时结束,当循环内的语句将控制转移到循环外或引发异常时也会结束。

语法格式如下。

```
while_loop_statement ::=WHILE boolean_expression
LOOP statement...
END LOOP [ label ] ;
```

while_loop_statement,boolean_expression 是值为 TRUE、FALSE 或 NULL 的表达式。

boolean_expression 在循环的每次迭代开始时进行评估。如果其值为 TRUE,则 LOOP 之后的语句将运行。否则,控制转移到 WHILE LOOP 语句之后的语句。

对 statement,为防止无限循环,至少有一条语句必须将 boolean_expression 的值更改为 FALSE 或 NULL,将控制转移到循环外或引发异常。

可以将控制转移到循环外的语句如下。

(1) CONTINUE 语句(当它将控制转移到封闭标记循环的下一次迭代时)。

(2) EXIT 语句。

(3) GOTO 语句。

(4) RAISE 语句。

label 是标识 while_loop_statement 的标签。CONTINUE、EXIT 和 GOTO 语句可以引用这个标签。

标签提高了可读性,尤其是当 LOOP 语句嵌套时,但前提是确保 END LOOP 语句中的标签与同一 LOOP 语句开头的标签匹配。

示例 2.16:使用 WHILE LOOP 语句输出 1~9,跳过 5。

程序代码如下。

```
DECLARE
  v_num number(10);
BEGIN
```

```
    v_num := 0;
    WHILE v_num < 10 LOOP
      EXIT WHEN v_num = 9;
      v_num = v_num + 1;
      IF v_num = 5 THEN
        CONTINUE ;
      END IF ;
      RAISE NOTICE '%', v_num;
    END LOOP ;
END
```

程序运行结果如下。

```
1
2
3
4
6
7
8
9
```

2.5.4　FOREACH 语句

FOREACH 循环类似 FOR 循环，但并不是通过一个 SQL 查询返回的行进行迭代，而是通过数组值的元素进行迭代。通常，FOREACH 通过一个组合表达式的部件加以迭代。

语法格式如下。

```
[ <<label>> ]
FOREACH target [ SLICE number ] IN ARRAY expression LOOP
    statements
END LOOP [ label ];
```

label 标签提高了可读性，尤其是当 LOOP 语句嵌套时，但前提是确保 END LOOP 语句中的标签与同一 LOOP 语句开头的标签匹配。

target 变量必须是一个数组，并且它接收数组值的连续切片，其中每一个切片都有 SLICE 指定的维度数。元素会被按照存储顺序访问，而不管数组的维度数。

给定一个正 SLICE 值，FOREACH 通过数组的切片而不是单一元素迭代。SLICE 值必须是一个不大于数组维度数的整数常量。

statements 语句中的 EXIT、EXIT WHEN、CONTINUE 或 CONTINUE WHEN 会导致循环或循环的当前迭代提前结束。

示例 2.17：本例使用 FOREACH 语句打印输出综合评价值大于等于 9 的商品名。

程序代码如下。

```
BEGIN
  FOR shops IN (SELECT * FROM "sales"."shopstores")
```

```
    LOOP
      IF shops.comprgrading >= 9 THEN
         RAISE NOTICE '商品名:%',shops.shopname ;
      ELSE
         NULL;
      END IF ;
   END LOOP ;
END
```

程序运行结果如下。

```
商品名:电子工业出版社
商品名:壹品宠物生活专营店
商品名:普安特旗舰店
商品名:蓝月亮京东自营旗舰店
商品名:小迷糊京东自营旗舰店
商品名:玖慕京东自营旗舰店
商品名:植迷者
商品名:genanx旗舰店
…
```

2.6 获取执行状态信息

2.6.1 获取结果状态和执行位置信息

1. 获取结果状态

有几种方法可以判断一条命令的效果。第一种方法是使用 GET DIAGNOSTICS 命令,其语法格式如下。

```
GET DIAGNOSTICS variable { = | := } item [ , ... ];
```

该命令允许检索系统状态指示符。每个 item 是一个关键字,它标识一个被赋予指定变量的状态值(变量应具有正确的数据类型来接收状态值)。表 2.1 展示了当前可用的状态项。

表 2.1 可用结果状态项

名称	类型	描述
ROW_COUNT	Bigint	最近的 SQL 命令影响的行数
SYS_CONTEXT	text	描述当前调用栈的文本行

第二种判断命令效果的方法是检查一个名为 FOUND 的 boolean 类型的特殊变量。在每一次 PL/SQL 函数调用时,FOUND 开始都为 false。它的值会被下面的每一种类型的语句设置。

① 如果一个 SELECT INTO 语句赋值了一行,它将把 FOUND 设置为 true,如果没有

返回行,则将其设置为 false。

② 如果一个 PERFORM 语句生成（或者抛弃）一行或多行,它将把 FOUND 设置为 true,如果没有产生行,则将其设置为 false。

③ 如果 UPDATE、INSERT 以及 DELETE 语句影响了至少一行,它们会把 FOUND 设置为 true,如果没有影响行,则将其设置为 false。

④ 如果一个 FETCH 语句返回了一行,它将把 FOUND 设置为 true,如果没有返回行,则将其设置为 false。

⑤ 如果一个 MOVE 语句成功地重定位了游标,它将会把 FOUND 设置为 true,否则设置为 false。

⑥ 如果一个 FOR 或 FOREACH 语句迭代了一次或多次,它将会把 FOUND 设置为 true,否则设置为 false。当循环退出时,FOUND 用这种方式设置；在循环执行中,尽管 FOUND 可能被循环体中其他语句的执行所改变,但它不会被循环语句修改。

⑦ 如果查询返回至少一行,RETURN QUERY 和 RETURN QUERY EXECUTE 语句会把 FOUND 设为 true,如果没有返回行,则设置为 false。

其他的 PL/SQL 语句不会改变 FOUND 的状态。尤其需要注意的是：EXECUTE 会修改 GET DIAGNOSTICS 的输出,但不会修改 FOUND 的输出。FOUND 是每个 PL/SQL 函数的局部变量；任何对它的修改只影响当前的函数。

示例 2.18：获取动态 SQL 影响的行数。

程序代码如下。

```
DECLARE
   affect_rows integer;
BEGIN
   EXECUTE 'UPDATE customers SET custname=custname';
   GET DIAGNOSTICS affect_rows = ROW_COUNT;
   RAISE NOTICE '%',affect_rows;
END;
```

程序运行结果如下。

```
NOTICE:4998
```

2. 获取执行位置信息

GET DIAGNOSTICS命令检索当前执行状态的信息(GET STACKED DIAGNOSTICS命令会把执行状态的信息报告成一个以前的错误)。它的 SYS_CONTEXT 状态项可用于标识当前执行位置。状态项 SYS_CONTEXT 将返回一个文本字符串,其中有描述该调用栈的多行文本。第一行会指向当前函数以及当前正在执行 GET DIAGNOSTICS 的命令。第二行及其后的行表示调用栈中更上层的调用函数。

示例 2.19：获取执行位置信息。

程序代码如下。

```
CREATE OR REPLACE FUNCTION outer_func() RETURNS integer AS
```

```
BEGIN
    RETURN inner_func();
END;

CREATE OR REPLACE FUNCTION inner_func() RETURNS integer AS
DECLARE
    stack text;
BEGIN
    GET DIAGNOSTICS stack = SYS_CONTEXT;
    RAISE NOTICE E'--- Call Stack ---\n%', stack;
    RETURN 1;
END;

SELECT outer_func();
```

程序运行结果如下。

```
--- Call Stack ---
PL/SQL 函数 inner_func()的第 5 行的 GET DIAGNOSTICS
PL/SQL 函数 outer_func()的第 3 行的 RETURN
outer_func|
----------+
        1|
```

2.6.2 错误和消息

PL/SQL 程序运行过程中不可避免地会出现错误,可以使用 RAISE 语句报告消息,并抛出错误。

```
RAISE [ level ] 'format' [, expression [, ... ]] [ USING option = expression [, ... ] ];
RAISE [ level ] condition_name [ USING option = expression [, ... ] ];
RAISE [ level ] SQLSTATE 'sqlstate' [ USING option = expression [, ... ] ];
RAISE [ level ] USING option = expression [, ... ];
RAISE ;
```

level 选项指定了错误的严重性。允许的级别有 DEBUG、LOG、INFO、NOTICE、WARNING 以及 EXCEPTION,默认级别是 EXCEPTION。EXCEPTION 会抛出一个错误(通常会中止当前事务)。其他级别仅仅是产生不同优先级的消息。一个特定优先级的消息无论被报告给客户端还是写到服务器日志,亦或是二者同时都做,都由 log_min_messages 和 client_min_messages 配置参数控制。

如果有 level,可以在其后写一个 format(必须是一个简单字符串而不是表达式)。该格式字符串指定被报告的错误消息文本。在格式字符串后面可以跟上可选的、被插入到该消息的参数表达式。在格式字符串中,%会被下一个可选参数的值所替换。%%可以发出一个字面的%。参数的数量必须匹配格式字符串中%占位符的数量,否则函数编译时会报错。

在 option=expression 项的后面加上 USING,可以为错误报告附加一些额外信息。每一个 expression 可以是任意字符串值的表达式。允许的 option 关键词是:

① MESSAGE：设置错误消息文本。该选项可以用于在 USING 之前包括一个格式字符串的 RAISE 形式。

② DETAIL：提供一个错误的细节消息。

③ HINT：提供一个提示消息。

④ ERRCODE：指定要报告的错误代码（SQLSTATE），可以用条件名，或者直接作为一个五字符的 SQLSTATE 代码。

⑤ COLUMN、CONSTRAINT、DATATYPE、TABLE、SCHEMA：提供一个相关对象的名称。

示例 2.20：使用 RAISE 语句报告消息以及抛出错误。

程序代码如下。

```
DECLARE
  v_job_id number(10) := 521;
  user_id number(10) := 1314;
BEGIN
--v_job_id 的值将替换字符串中的%
  RAISE NOTICE 'Calling cs_create_job(%)', v_job_id;
--用给定的错误消息和提示中止事务：
  RAISE EXCEPTION 'Nonexistent ID --> %', user_id
  USING HINT = 'Please check your user ID';
END
```

程序运行结果如下。

```
Calling cs_create_job(521)
SQL 错误 [P0001]: 错误: Nonexistent ID --> 1314
  Hint: Please check your user ID
  Where: PL/SQL 函数 inline_code_block 的第 8 行的 RAISE
```

除了上述错误消息外，还存在其他异常情况导致的错误信息，请参考第 7 章中关于异常信息的捕获内容描述。

第 3 章

PL/SQL 的复合数据类型

在 PL/SQL 数据类型中,复合数据类型包括集合类型和记录类型。其中,集合类型是多个相同类型的分量的集合,类似于高级语言中的数组。而记录类型是多个不同类型的变量的集合,类似于高级语言中的结构体类型。通常,在 PL/SQL 程序中,集合对应于表中某一列的多个值的集合,而记录类型对应一条记录中若干个字段的集合。本章将主要介绍以下内容。

- 集合类型。
- 记录类型。

 ## 3.1 集合类型

在集合类型中,内部元素始终具有相同的数据类型,可以通过集合变量的唯一索引访问它的每个元素。语法格式如下。

```
variable_name(index)
```

创建集合变量,需要先定义集合类型,然后创建该类型的变量,或者使用%TYPE 创建。
PL/SQL 有 3 种集合类型。
(1) 关联数组(Associative Array)。
(2) 可变数组(Varray)。
(3) 嵌套表(Nested Table)。
表 3.1 对 3 种集合类型进行了比较。

表 3.1 三种集合类型比较

集合类型	元素数量	索引类型	密度	未初始化状态	何处定义	作为复合类型的元素
关联数组	不指定	String or PLS_INTEGER	不定	空	PL/SQL 块或包	否
可变数组	指定	INTEGER	密集	NULL	PL/SQL 块、包或模式	模式下定义可以
嵌套表	不指定	INTEGER	开始密集,可变为稀疏	NULL	PL/SQL 块、包或模式	模式下定义可以

其中：

① 元素数量。如果指定了元素数量，则该数量为集合中的最大元素数。如果未指定元素数，则集合中的最大元素数为索引类型的数量上限。

② 密度。密集集合的元素之间没有间隙，第一个和最后一个元素之间的每个元素都必须已定义并具有一个值（该值可以为 NULL，除非该元素具有 NOT NULL 约束）。稀疏集合的元素之间可以存在间隙。

③ 未初始化状态。空集合，一个没有元素的集合。要添加元素，需要调用集合的 EXTEND 方法；NULL 集合，一个不存在的集合。要将 NULL 集合更改为存在的状态，必须通过将其设置为空或为其分配非空值来初始化它（通过调用相关构造函数）。无法通过直接调用 EXTEND 方法来初始化一个 NULL 集合。

④ 何处定义。PL/SQL 块中定义的集合类型属于本地类型。它仅支持在块中使用，并且仅当块位于独立子程序或包子程序中时才会被存储在数据库中。包规范中定义的集合类型是公共类型。可以通过包名限定（package_name.type_name）从包外部引用它。它会一直存储在数据库中，直到删除包。在模式中定义的集合类型是独立类型，可以使用 CREATE TYPE 创建它。它存储在数据库中，直到使用 DROP TYPE 删除它。

⑤ 作为复合类型的元素：要成为复合类型的元素类型，集合类型必须是独立集合类型，即定义在模式中的集合类型。

3.1.1 关联数组

关联数组是一组键值对。每个键都是一个唯一的索引，用于定位与之相关联的值，语法格式如下。

```
variable_name(index)
```

索引的数据类型可以是字符串类型（VARCHAR2、VARCHAR、STRING 或 LONG）或 PLS_INTEGER。其中数据是按索引排序顺序存储，而不是按创建顺序存储。

与数据库表相同的是：

① 关联数组在填充之前为空，但不为 NULL。

② 关联数组可以容纳不定量的元素，可以在不知道其位置的情况下访问这些元素。

与数据库表不同的是：

① 关联数组不需要磁盘空间或网络操作。

② 关联数组不能使用 DML 语句操作。

1. 关联数组的声明

示例 3.1：以字符串为索引的关联数组。

功能描述：本例定义一种按字符串索引的关联数组，并声明该类型的变量，用 3 个元素填充该关联数组，然后更改其中一个元素的值，并打印值（数据是按索引排序顺序存储，而不是按创建顺序）。

程序代码如下。

```
DECLARE
```

```
--Associative array indexed by string:
TYPE ass_type IS TABLE OF INT
INDEX BY VARCHAR2(64);
age ass_type;
i VARCHAR2(64);

BEGIN
  --Add elements (key-value pairs) to associative array:
  age('zs') := 20;
  age('ls') := 30;
  age('ww') := 40;
  --Change value associated with key 'Smallville':
  age('zs') := 25;
  --Print associative array:
  i := age.FIRST;
  WHILE i IS NOT NULL LOOP
      RAISE NOTICE '% is % years old', i, age(i);
      i := age.NEXT(i);
  END LOOP;
END;
```

程序运行结果如下。

```
NOTICE:ls is 30 years old
NOTICE:ww is 40 years old
NOTICE:zs is 25 years old
```

示例 3.2：函数返回以 PLS_INTEGER 为索引的关联数组。

功能描述：本例定义了一种以 PLS_INTEGER 为索引的关联数组，以及一个返回该关联数组类型的函数。

程序代码如下。

```
DECLARE
  TYPE area_of_circle IS TABLE OF NUMBER INDEX BY PLS_INTEGER;
  area area_of_circle;      --result variable
  num INT = 3;
  FUNCTION get_area_of_circle (
      num INT
  ) RETURN area_of_circle
  IS
      s area_of_circle;
  BEGIN
      FOR i IN 1..num LOOP
          s(i) := 2 * 3.14 * i * i;
      END LOOP;
      RETURN s;
  END;
BEGIN
  area= get_area_of_circle (num);
```

```
    FOR i in 1..num LOOP
        RAISE NOTICE 'area(%) = %', i, area(i);
    END LOOP;
END;
```

程序运行结果如下。

```
NOTICE:  area(1) = 6.28
NOTICE:  area(2) = 25.12
NOTICE:  area(3) = 56.52
```

2. 关联数组的适用情况

关联数组适用于以下情况。

① 查找一个相对较小的表,每次调用子程序或初始化声明子程序的包时,都可以在内存中构造它。

② 向数据库服务器传递集合和从数据库服务器传递集合。

关联数组用于临时数据存储。要使关联数组在数据库会话的生命周期内保持存在,请在包规范中声明它,并在包体中对其进行填充赋值。

3.1.2 可变数组

可变数组是一个数组,其元素数为从零(空)到声明的最大值之间,大小不等。

访问可变数组变量的元素,可以使用语法 variable_name(index)。指数的下限为 1,上限是当前元素的数量。上限在添加或删除元素时会发生变化,但不能超过声明时指定的最大值。从数据库中存储和检索可变数组时,其索引和元素顺序保持对应。

未初始化的可变数组变量是一个空集合。必须通过构造函数或者为其赋予一个非空的值来初始化它。

1. 可变数组的声明

示例 3.3:可变数组。

功能描述:本例定义了一个本地可变数组类型,声明了该类型的变量,并使用构造函数进行初始化,定义了一个打印可变数组的存储过程。之后调用该存储过程 3 次:分别为初始化变量后,更改两个元素的值后,以及使用构造函数更改所有元素的值后。

程序代码如下。

```
DECLARE
  --VARRAY type
  TYPE VARRAY_TYPE IS VARRAY(4) OF VARCHAR2(15);
  --varray variable initialized with constructor:
  class VARRAY_TYPE := VARRAY_TYPE('zs', 'ls', 'ww', 'zl');

  PROCEDURE show_class (heading VARCHAR2) IS
  BEGIN
      RAISE NOTICE '%', heading;
```

```
        FOR i IN 1..4 LOOP
            RAISE NOTICE '%.%',i,class(i);
        END LOOP;
        RAISE NOTICE '---';
    END;

BEGIN
    show_class('2001 Class:');
    --Change values of two elements
    class(3) := 'xx';
    class(4) := 'xm';
    show_class('2005 Class:');

    --Invoke constructor to assign new values to varray variable:
    class := VARRAY_TYPE('xz', 'xw', 'xc', 'xx');
    show_class('2009 Class:');
END;
```

程序运行结果如下。

```
NOTICE:2001 Class:
NOTICE:1.zs
NOTICE:2.ls
NOTICE:3.ww
NOTICE:4.zl
NOTICE:---
NOTICE:2005 Class:
NOTICE:1.zs
NOTICE:2.ls
NOTICE:3.xx
NOTICE:4.xm
NOTICE:---
NOTICE:2009 Class:
NOTICE:1.xz
NOTICE:2.xw
NOTICE:3.xc
NOTICE:4.xx
NOTICE:---
```

2. 可变数组的适用情况

可变数组适用于以下情况。

① 已知元素的最大数量。

② 需要按顺序访问元素。

因为必须同时存储或检索所有元素,所以,可变数组一般不适用于拥有大量元素的情况。

3.1.3 嵌套表

在数据库中,嵌套表是一种可以不指定顺序来存储未指定数量行的类型。

从数据库中检索嵌套表值到 PL/SQL 嵌套表变量时，PL/SQL 会从 1 开始为行提供连续索引。使用这些索引，可以访问嵌套表变量的各个行。语法是 variable_name(index)。从数据库中存储和检索嵌套表时，嵌套表的索引和行顺序可能不稳定。

当添加或删除元素时，嵌套表变量占用的内存量可以动态地增加或减少。

未初始化的嵌套表变量是一个 NULL 集合。必须通过构造函数或为其赋予非空值进行初始化。

1. 嵌套表的声明

示例 3.4：块内部的嵌套表类型。

功能描述：本例定义一个块内部的嵌套表类型，然后声明该类型的变量（使用构造函数初始化），并定义一个打印嵌套表的存储过程。之后调用该存储过程 3 次：初始化变量后、更改一个元素的值后，以及使用构造函数更改所有元素的值后。在第 2 次构造函数调用之后，嵌套表只有 2 个元素，引用元素 3 会引发错误。

程序代码如下。

```
DECLARE
  --nested table type
  TYPE NESTTAB_TYPE IS TABLE OF VARCHAR2(15);
  --nested table variable initialized with constructor:
  fruits NESTTAB_TYPE := NESTTAB_TYPE('Apple', 'Orange', 'Banana', 'PEAR');

  PROCEDURE show_fruits (heading VARCHAR2) IS
  BEGIN
      RAISE NOTICE '%', heading;

      FOR i IN fruits.FIRST .. fruits.LAST LOOP
          RAISE NOTICE '%', fruits(i);
      END LOOP;

      RAISE NOTICE '---';
  END;

BEGIN
  show_fruits('Initial Values:');
  --Change value of one element
  fruits(3) := 'Watermelon';
  show_fruits('Current Values:');
  --Change entire table
  fruits := NESTTAB_TYPE('Strawberry', 'Pineapple');
  show_fruits('Current Values:');
END;
```

程序运行结果如下。

```
NOTICE:   Initial Values:
NOTICE:   Apple
NOTICE:   Orange
```

```
NOTICE:  Banana
NOTICE:  PEAR
NOTICE:  ---
NOTICE:  Current Values:
NOTICE:  Apple
NOTICE:  Orange
NOTICE:  Watermelon
NOTICE:  PEAR
NOTICE:  ---
NOTICE:  Current Values:
NOTICE:  Strawberry
NOTICE:  Pineapple
NOTICE:  ---
```

示例 3.5：独立的嵌套表类型。

功能描述：本例定义了一个独立的类型 nest_type 和 1 个独立的存储过程 show_nesttype，用来打印该类型的变量。匿名块声明 1 个类型为 nest_type 的嵌套表变量，用构造函数将其初始化为空，变量初始化后，使用构造函数更改所有元素的值，两次调用 show_nesttype。

程序代码如下。

```
CREATE OR REPLACE TYPE nest_type IS TABLE OF NUMBER;

CREATE OR REPLACE PROCEDURE show_nesttype (nt nest_type) AUTHID DEFINER IS
  i  NUMBER;
BEGIN
  i := nt.FIRST;

  IF i IS NULL THEN
     RAISE NOTICE 'nest type is empty';
  ELSE
     WHILE i IS NOT NULL LOOP
        RAISE NOTICE 'nt(%) = %', i, nt(i);
        i := nt.NEXT(i);
     END LOOP;
  END IF;

  RAISE NOTICE '---';
END show_nesttype;

DECLARE
   nt nest_type := nest_type();      --nested table variable initialized to empty
BEGIN
   show_nesttype(nt);
   nt := nest_type(1,3,99,1001);
   show_nesttype(nt);
END;
```

程序运行结果如下。

```
NOTICE:  nest type is empty
NOTICE:  ---
NOTICE:  nt(1) = 1
NOTICE:  nt(2) = 3
NOTICE:  nt(3) = 99
NOTICE:  nt(4) = 1001
NOTICE:  ---
```

2. 嵌套表的适用情况

嵌套表适用于以下情况。

① 未指定元素的数量。

② 索引值是不连续的。

③ 需要删除或更新某些元素,但不能同时删除或更新所有元素。

3.1.4 集合的构造函数

集合的构造函数(constructor)是一个系统定义的函数,与集合类型同名,返回值为对应集合类型。

构造函数调用的语法格式如下。

```
collection_type ( [ value [, value ]... ] )
```

如果参数列表为空,则构造函数返回一个空集合。否则,构造函数将返回包含指定值的集合。

可以在变量声明和块的执行部分中将返回的集合分配给相同类型的集合变量。

示例 3.6:将可变数组变量初始化为空。

功能描述:本例调用了一个构造函数 2 次,将可变数组变量 class 在声明中初始化为空,并在块的执行部分为其提供新值。应用存储过程 show_class 打印变量 class 的元素值。为了确定集合何时为空,show_class 使用了集合的 COUNT 方法。

程序代码如下。

```
DECLARE
  TYPE VARRAY_TYPE IS VARRAY(4) OF VARCHAR2(15);
  class VARRAY_TYPE := VARRAY_TYPE();    --initialize to empty

  PROCEDURE show_class (heading VARCHAR2)
  IS
  BEGIN
     RAISE NOTICE '%', heading;

     IF class.COUNT = 0 THEN
        RAISE NOTICE 'Empty';
     ELSE
```

```
            FOR i IN 1..4 LOOP
               RAISE NOTICE '%.%',i,class(i);
            END LOOP;
         END IF;

       RAISE NOTICE '---';
     END;
BEGIN
    show_class('Class:');
    class := VARRAY_TYPE('xx', 'xm', 'xz', 'xh');
    show_class('Class:');
END;
```

程序运行结果如下。

```
NOTICE:  Class:
NOTICE:  Empty
NOTICE:  ---
NOTICE:  Class:
NOTICE:  1.xx
NOTICE:  2.xm
NOTICE:  3.xz
NOTICE:  4.xh
NOTICE:  ---
```

3.1.5 集合变量赋值

可以通过以下方式为集合变量赋值。

① 调用构造函数来创建集合,并将其分配给集合变量。
② 使用赋值语句将另一个现有集合变量的值赋值给它。
③ 将其作为 OUT 或 IN OUT 参数传递给子程序,然后在子程序内赋值。

要为集合变量的标量元素赋值,使用 collection_variable_name(index)语法引用这些元素,并为其赋值。

1. 数据类型兼容

只有当集合变量具有相同的数据类型时,才能将集合分配给集合变量。只有元素类型相同,则无法互相赋值。

示例 3.7:数据类型兼容的集合赋值。

功能描述:在本例中,可变数组类型 VARRAY_TYPE1 和 VARRAY_TYPE2 有相同的元素类型 TEXT。集合变量 varray1 和 varray2 有相同的数据类型 VARRAY_TYPE1,但是集合变量 varray3 是数据类型 VARRAY_TYPE2。varray1 给 varray2 赋值成功,但 varray1 给 varray3 赋值失败。

程序代码如下。

```
DECLARE
  TYPE VARRAY_TYPE1 IS VARRAY(5) OF TEXT;
```

```
    TYPE VARRAY_TYPE2 IS VARRAY(5) OF TEXT;

  varray1 VARRAY_TYPE1 := VARRAY_TYPE1('Jones','Wong','Marceau');
  varray2 VARRAY_TYPE1;
  varray3 VARRAY_TYPE2;
BEGIN
  varray2 := varray1;   --ok
  varray3 := varray1;   --error
END;
```

程序运行结果如下。

```
ERROR:  can not cast from Collection type to Collection type.
```

2. 给可变数组和嵌套表变量赋 NULL 值

可以给可变数组或嵌套表变量赋 NULL 值或相同数据类型的 NULL 集合。任一赋值都会使变量为空。

示例 3.8：给嵌套表变量赋 NULL 值。

功能描述：在本例中，先将嵌套表变量 city_names 初始化为非空值；然后为其分配空集合置空；最后将其重新初始化为不同的非空值。

程序代码如下。

```
DECLARE
  TYPE cnames_tab IS TABLE OF VARCHAR2(30);
  --Initialized to non-null value
  city_names cnames_tab := cnames_tab(
    'Beijing','Shanghai','Guangzhou','Shenzhen');
  --Not initialized, therefore null
  empty_set cnames_tab;

  PROCEDURE show_city_names_status IS
  BEGIN
     IF city_names IS NULL THEN
         RAISE NOTICE 'city_names is null.';
     ELSE
         RAISE NOTICE 'city_names is not null.';
     END IF;
  END  show_city_names_status;

BEGIN
  show_city_names_status;
  city_names := empty_set;   --Assign null collection to city_names.
  show_city_names_status;
  city_names := cnames_tab (
    'Shipping','Sales','Finance','Payroll');  --Re-initialize city_names
  show_city_names_status;
END;
```

程序运行结果如下。

```
NOTICE:  city_names is not null.
NOTICE:  city_names is null.
NOTICE:  city_names is not null.
```

3.1.6 多维集合

集合只有一个维度,但可以通过使集合的元素为另一个集合来构造一个多维集合。

示例 3.9:二维可变数组。

功能描述:在本例中,nva 是一个二维可变数组(元素为整型可变数组的数组)。

程序代码如下。

```
DECLARE
  TYPE v1 IS VARRAY(10) OF INTEGER;      --varray of integer
  nva v1 := v1(2,3,5);
  TYPE v2 IS VARRAY(10) OF v1;           --varray of varray of integer
  nva v2 := v2(va, v1(1,2,3), v1(4,5), va);
  i INTEGER;
  val v1;
BEGIN
  i := nva(2)(3);
  RAISE NOTICE 'i = %', i;
  nva.EXTEND;
  nva(5) := v1(6, 7);                    --replace inner varray elements
  nva(4) := v1(8,9,10,10000);            --replace an inner integer element
  nva(4)(4) := 1;                        --replace 43345 with 1
  nva(4).EXTEND;                         --add element to 4th varray element
  nva(4)(5) := 98;                       --store integer 89 there
END;
```

程序运行结果如下。

```
NOTICE:   i = 3
```

示例 3.10:嵌套表类型的嵌套表及整型可变数组的嵌套表。

功能描述:在本例中,ntb1 是元素为字符串嵌套表的嵌套表,ntb2 是元素为整型可变数组的嵌套表。

程序代码如下。

```
DECLARE
  TYPE tab1 IS TABLE OF VARCHAR2(20);    --nested table of strings
  vtb1 tab1 := tab1('t1', 't2');
  TYPE ntab1 IS TABLE OF tab1;           --nested table of nested tables of strings
  vntb1 ntab1 := ntab1(vtb1);
  TYPE tv1 IS VARRAY(10) OF INTEGER;     --varray of integers
  TYPE ntb2 IS TABLE OF tv1;             --nested table of varrays of integers
```

```
    vntb2 ntb2 := ntb2(tv1(1,2), tv1(3,4,5));
BEGIN
  vntb1.EXTEND;
  vntb1(2) := vntb1(1);
  vntb1.DELETE(1);                    --delete first element of vntb1
  vntb1(2).DELETE(1);    --delete first string from second table in nested table
END;
```

3.1.7 集合的比较

要确定一个集合变量与另一个集合变量的大小,必须定义在该内容中小于的含义,并编写一个返回 TRUE 或 FALSE 的函数。

嵌套表除了相等和不相等外,无法使用其他关系运算符对两个集合变量进行比较。这种限制也适用于隐式比较。例如,集合变量不能出现在 DISTINCT、GROUP BY 或 ORDER BY 子句中。

1. 将可变数组和嵌套表变量与 NULL 进行比较

可以使用 IS [NOT] NULL 运算符,将可变数组和嵌套表变量与 NULL 进行比较,但是不能使用关系运算符等于(=)和不等于(<>,!= 或^=)进行比较。

示例 3.11:将可变数组和嵌套表变量与 NULL 进行比较。

程序代码如下。

```
DECLARE
  TYPE VARRAY_TYPE IS VARRAY(4) OF VARCHAR2(15);    --VARRAY type
  class VARRAY_TYPE;                                --varray variable
  TYPE NEST_TYPE IS TABLE OF TEXT;                  --nested table type
  fruits NEST_TYPE := NEST_TYPE('Pear', 'Apple');   --nested table variable

BEGIN
  IF class IS NULL THEN
    RAISE NOTICE 'class IS NULL';
  ELSE
    RAISE NOTICE 'class IS NOT NULL';
  END IF;

  IF fruits IS NOT NULL THEN
    RAISE NOTICE 'fruits IS NOT NULL';
  ELSE
    RAISE NOTICE 'fruits IS NULL';
  END IF;
END;
```

程序结果运行如下。

```
NOTICE:  class IS NULL
NOTICE:  fruits IS NOT NULL
```

2. 嵌套表的相等和不相等比较

两个嵌套表变量只有在拥有相同元素集的情况下才相等（任何顺序都可以）。

如果两个嵌套表变量具有相同的嵌套表类型，并且该嵌套表类型没有记录类型的元素，则可以使用关系运算符比较这两个变量的相等（=）或不相等（<>，!=，^=）。

示例 3.12：嵌套表的相等和不相等比较。

程序代码如下。

```
DECLARE
  TYPE NEST_TYPE IS TABLE OF VARCHAR2(30); -- element type is not record type
  fruits1 NEST_TYPE :=
    NEST_TYPE('Apple','Banana','Pear','Orange');
  fruits2 NEST_TYPE :=
    NEST_TYPE('Apple','Pear','Banana','Orange');
  fruits3 NEST_TYPE :=
    NEST_TYPE('Apple','Pineapple','Strawberry');
BEGIN
  IF fruits1 = fruits2 THEN
    RAISE NOTICE 'dept_names1 = dept_names2';
  END IF;
  IF fruits2 != fruits3 THEN
    RAISE NOTICE 'dept_names2 != dept_names3';
  END IF;
END;
```

程序运行结果如下。

```
NOTICE:  dept_names1 = dept_names2
NOTICE:  dept_names2 != dept_names3
```

3.1.8 集合方法

集合方法是 PL/SQL 子程序，既可以是一个返回集合信息的函数，也可以是一个对集合进行操作的过程。集合方法使集合更易于使用和维护。

表 3.2 描述了各种集合方法的情况。

表 3.2 各种集合方法

方　　法	类　　型	描　　述
DELETE	存储过程	从集合中删除元素
TRIM	存储过程	从可变数组或嵌套表的末尾删除元素
EXTEND	存储过程	从可变数组或嵌套表的末尾增加元素
FIRST	函数	返回集合的第一个索引
LAST	函数	返回集合的最后一个索引
COUNT	函数	返回集合元素的数量

续表

方法	类型	描述
LIMIT	函数	返回集合可以包含的最大的元素数量
PRIOR	函数	返回指定索引之前的索引
NEXT	函数	返回指定索引的下一个索引

集合方法的基本调用语法格式如下。

```
collection_name.method
```

集合方法可以出现在任何调用 PL/SQL 子程序(函数或过程)的地方,SQL 语句除外。在子程序中,集合作为参数程序的参数继承了对应集合的属性。可以将集合方法应用于此类参数。对于 varray 数组参数,无论参数模式如何,LIMIT 的值总是源于参数类型定义。

示例 3.13:不同集合类型方法的调用。

程序代码如下。

```
DECLARE
    TYPE ASS_TYPE IS TABLE of int index by VARCHAR2(15);    --关联数组
    TYPE VARRAY_TYPE IS VARRAY(5) OF int;                   --可变数组
    TYPE NESTTAB_TYPE IS TABLE OF int;                      --嵌套表
    v_indextab ASS_TYPE;                                    --声明关联数组类型的变量
    v_array VARRAY_TYPE := VARRAY_TYPE(10,20);              --声明可变数组变量
    v_nesttab NESTTAB_TYPE := NESTTAB_TYPE(10,20,30);       --声明嵌套表类型变量

    --输出关联数组
    PROCEDURE SHOW_ASSTYPE(at ASS_TYPE) IS
      i TEXT;
    BEGIN
      i := at.FIRST;
      IF i IS NULL THEN
        DBMS_OUTPUT.PUT_LINE('asstype is empty');
      ELSE
        WHILE i IS NOT NULL LOOP
          RAISE NOTICE 'at.(%) = %', i, at(i);
          i := at.NEXT(i);
        END LOOP;
      END IF;

        RAISE NOTICE '---';
    END;

    --输出可变数组
    PROCEDURE SHOW_ARRAY(ay VARRAY_TYPE)
      IS
      BEGIN
        IF ay IS NULL THEN
```

```
        RAISE NOTICE 'Does not exist';
      ELSIF ay.FIRST IS NULL THEN
        RAISE NOTICE 'Has no members';
      ELSE
        FOR i IN ay.FIRST..ay.LAST LOOP
          RAISE NOTICE '%.%', i, ay(i);
        END LOOP;
      END IF;

      RAISE NOTICE '---';
    END;

  --输出嵌套表
  PROCEDURE SHOW_NESTTYPE (nt NESTTAB_TYPE) IS
    i NUMBER;
    BEGIN
    i:= nt.FIRST;
      IF i IS NULL THEN
        RAISE NOTICE 'nest type is empty';
      ELSE
        WHILE i IS NOT NULL LOOP
          RAISE NOTICE 'nt(%) = %', i, nt(i);
          i := nt.NEXT(i);
        END LOOP;
      END IF;
      RAISE NOTICE '---';
  END;
BEGIN
  RAISE NOTICE '=============关联数组=============';
  v_indextab('A') := 10;
  v_indextab('B') := 20;
  v_indextab('C') := 30;
  v_indextab('D') := 40;
  v_indextab('E') := 50;
  SHOW_ASSTYPE(v_indextab);
  v_indextab.DELETE('C');
  SHOW_ASSTYPE(v_indextab);
  v_indextab.DELETE('A','D');
  SHOW_ASSTYPE(v_indextab);
  v_indextab.DELETE;
  SHOW_ASSTYPE(v_indextab);

  RAISE NOTICE '=============可变数组=============';
  v_array.EXTEND(3,1);
  SHOW_ARRAY(v_array);
  v_array.TRIM(2);
  SHOW_ARRAY(v_array);

  RAISE NOTICE '=============嵌套表=============';
  v_nesttab.EXTEND(2,1);
```

```
    SHOW_NESTTYPE(v_nesttab);
    v_nesttab.DELETE(5);
    SHOW_NESTTYPE(v_nesttab);
    v_nesttab.EXTEND;
    SHOW_NESTTYPE(v_nesttab);
    v_nesttab.TRIM;
    SHOW_NESTTYPE(v_nesttab);
END;
```

程序运行结果如下。

```
NOTICE:=============关联数组=============
NOTICE:at.(A) = 10
NOTICE:at.(B) = 20
NOTICE:at.(C) = 30
NOTICE:at.(D) = 40
NOTICE:at.(E) = 50
NOTICE:---
NOTICE:at.(A) = 10
NOTICE:at.(B) = 20
NOTICE:at.(D) = 40
NOTICE:at.(E) = 50
NOTICE:---
NOTICE:at.(E) = 50
NOTICE:---
NOTICE:---
NOTICE:=============可变数组=============
NOTICE:1.10
NOTICE:2.20
NOTICE:3.10
NOTICE:4.10
NOTICE:5.10
NOTICE:---
NOTICE:1.10
NOTICE:2.20
NOTICE:3.10
NOTICE:---
NOTICE:=============嵌套表=============
NOTICE:nt(1) = 10
NOTICE:nt(2) = 20
NOTICE:nt(3) = 30
NOTICE:nt(4) = 10
NOTICE:nt(5) = 10
NOTICE:---
NOTICE:nt(1) = 10
NOTICE:nt(2) = 20
NOTICE:nt(3) = 30
NOTICE:nt(4) = 10
NOTICE:---
```

```
NOTICE:nt(1) = 10
NOTICE:nt(2) = 20
NOTICE:nt(3) = 30
NOTICE:nt(4) = 10
NOTICE:nt(6) = <NULL>
NOTICE:---
NOTICE:nt(1) = 10
NOTICE:nt(2) = 20
NOTICE:nt(3) = 30
NOTICE:nt(4) = 10
NOTICE:---
```

3.2 记录类型

3.2.1 记录类型概述

在记录类型中，内部元素可以有不同的数据类型，每一个元素称为字段。可以通过字段名称访问记录变量的每个字段，语法是 variable_name.field_name。可以把记录类型想象成面向对象中的类。

3.2.2 声明记录类型

可以通过以下 3 种方式创建记录类型变量。

（1）使用 RECORD 语句自定义记录类型，然后使用该记录类型声明记录类型变量。

在 PL/SQL 块中定义的 RECORD 类型是本地类型，仅在块中有效。

在包规范中定义的 RECORD 类型是公共项。可以通过包名限定的方式（package_name.type_name）在包外引用。它会一直存储在数据库中，直到通过 DROP PACKAGE 语句删除包。

不能在模式中创建 RECORD 类型，因此，RECORD 类型不可以作为 ADT 属性数据类型。

为了定义记录类型，需要指定名称和定义字段，字段默认值为 NULL。可以给字段加 NOT NULL 约束，这种情况下必须指定一个非 NULL 的初始值。如果没有 NOT NULL 约束，则这个非 NULL 初始值可选。

（2）通过%ROWTYPE 声明一个记录类型，该变量可以表示数据库表或视图的完整行或部分行。

%ROWTYPE 属性允许声明一个 record 变量，该变量表示数据库表或视图的完整行或部分行。要声明始终代表数据库表或视图的整行 record 变量，语法格式如下。

```
variable_name table_or_view_name%ROWTYPE;
```

对于表或视图的每一列，record 变量都有一个相同名称和数据类型的字段。

（3）通过%TYPE 属性，可以声明与之前声明的变量或列具有相同数据类型的数据项（而不需要知道该类型具体是什么）。

3.2.3 使用记录类型

声明记录类型后,可以用来定义记录类型变量。以下各示例演示各记录类型如何使用。

示例 3.14：记录类型定义和变量声明。

程序代码如下。

```
DECLARE
  TYPE StuRecTyp IS RECORD (
    stu_id    NUMBER(4) NOT NULL := 1,
    stu_name  VARCHAR2(30) NOT NULL := 'xm',
    score     NUMBER(5,2) = 88.88
  );

stu_rec StuRecTyp;
BEGIN
  RAISE NOTICE 'stu_id = %', stu_rec.stu_id;
  RAISE NOTICE 'stu_name = %', stu_rec.stu_name;
  RAISE NOTICE 'score = %', stu_rec.score;
END;
```

程序运行结果如下。

```
NOTICE:   stu_id = 1
NOTICE:   stu_name = xm
NOTICE:   score = 88.88
```

示例 3.15：声明 RECORD(指定默认值)。

程序代码如下。

```
DECLARE
  TYPE Rec IS RECORD (a NUMBER, b NUMBER);
  r Rec := (0,1);
BEGIN
  raise notice 'r = %',r;
  raise notice 'r.a = %',r.a;
  raise notice 'r.b = %',r.b;
END;
```

程序运行结果如下。

```
NOTICE:   r = (0,1)
NOTICE:   r.a = 0
NOTICE:   r.b = 1
```

示例 3.16：嵌套的记录类型。

功能描述：本例定义了 2 个记录类型：stu_info_rec 和 stu_rec。类型 stu_rec 有 1 个字段为 stu_info_rec 类型。

程序代码如下。

```
DECLARE
  TYPE stu_info_rec IS RECORD (
    id NUMBER(4),
    name VARCHAR2(30)
  );

  TYPE stu_rec IS RECORD (
    stu_info   stu_info_rec,        --nested record
    score NUMBER(5,2)
  );

  student1 stu_rec;
BEGIN
  student1.stu_info.id := 1;
  student1.stu_info.name := 'xx';
  student1.score := 88;

  RAISE NOTICE '%, %, %', student1.stu_info.id, student1.stu_info.name, student1.score;
END;
```

程序运行结果如下。

```
NOTICE:  1, xx, 88
```

示例 3.17：包含 Varray 数组的记录类型。

功能描述：本例定义了 Varray 数组类型 stu_info_rec 和记录类型 stu_rec。类型 stu_rec 有 1 个字段为 stu_info_rec 类型。

程序代码如下。

```
DECLARE
  TYPE stu_info_rec IS VARRAY(2) OF VARCHAR2(20);
  TYPE stu_rec IS RECORD (
    stu_info   stu_info_rec := stu_info_rec('ay','fixd'),
    score NUMBER(5,2)
  );

  student1 stu_rec;
BEGIN
  student1.score := 88;

  RAISE NOTICE '%, %, %', student1.stu_info(1), student1.stu_info(2), student1.score;
END;
```

程序运行结果如下。

```
NOTICE:  1, xx, 88
```

示例 3.18：%ROWTYPE 变量表示完整的数据库表行。

功能描述：本例声明了 1 个 record 变量，该变量表示表 customers 的 1 行，为其字段赋值，并打印。

程序代码如下。

```
DECLARE
  cust_rec customers%ROWTYPE;
BEGIN
  --Assign values to fields:
  cust_rec.custid   := 1;
  cust_rec.custname := 'zs';
  cust_rec.gender := '男';

  --Print fields:
  RAISE NOTICE 'cust_rec.custid = %, cust_rec.custname = %, cust_rec.gender = %
', cust_rec.custid, cust_rec.custname, cust_rec.gender;
END;
```

程序运行结果如下。

```
NOTICE:cust_rec.custid = 1, cust_rec.custname = zs, cust_rec.gender = 男
```

第 4 章

PL/SQL 中的静态 SQL 语句

从编译和运行的角度,SQL 语句可以分为静态 SQL 和动态 SQL。在包含 SQL 语句的代码块被编译时,如果这个 SQL 语句是有明确定义的,则称该 SQL 语句是静态 SQL;而动态 SQL 语句是在应用程序运行时被编译和执行的。本章将主要介绍 PL/SQL 中的静态 SQL 语句及其应用,包括以下内容。

- 静态 SQL 语句概述。
- 游标。
- 游标变量。
- 批量处理。

 ## 4.1 静态 SQL 语句概述

4.1.1 静态 SQL 语句类型

静态 SQL 语句可以直接在 PL/SQL 中使用,其语法与标准 SQL 语句完全一致,除非另有说明。PL/SQL 中允许出现以下类型的 SQL 语句。

(1) 查询语句(SELECT)。

从一个或更多表中返回记录行。

(2) 数据定义语言(DDL)语句。

① CREATE 在数据库中创建一个新的对象。

② ALTER 修改对象的属性。

③ DROP 从数据库中删除一个对象。

④ TRUNCATE 删除表中的数据。

(3) 数据操作语言(DML)语句。

① INSERT 将一个或多个新的行插入到表中。

② UPDATE 更新表中的一个或多个行中的一个或多个列的值。

③ DELETE 从表中删除一个或多行。

④ MERGE 非声明式的"更新插入"。如果表中已经存在指定列值的一行记录,则对它进行更新,否则进行插入。

（4）事务控制语言（TCL）语句。

① COMMIT 提交自上次 COMMIT 或 ROLLBACK 以来所有未完成的变化，并释放被锁的资源。

② ROLLBACK 撤销自指定的保存点成立以来所有未完成的变化，并释放这一部分代码使用的被锁的资源。

③ SET TRANSACTION 语句用于设置事务的只读、读/写、隔离级别以及分配名称等操作。

（5）表锁定语句（LOCK TABLE）。

允许在指定模式下锁定整个数据库表。这个命令覆盖了默认的通常应用于表的行级锁。

4.1.2 PL/SQL 中的 SELECT 语句

与标准的 SELECT 语句不同，在 PL/SQL 程序中，SELECT 语句需要与 INTO 或 BULK COLLECT INTO 短语结合使用，将查询结果保存到变量中。

1. SELECT INTO 语句

SELECT INTO 语句只能查询一条记录的信息，如果没有查询到任何数据，会产生 NO_DATA_FOUND 异常；如果查询到多条记录，则会产生 TOO_MANY_ROWS 异常。

INTO 句子后的变量用于接收查询的结果，变量的个数、顺序应该与查询的目标数据相匹配，也可以是记录类型的变量。

示例 4.1：SELECT INTO 查询语句。

功能描述：本示例根据具体订单号查询承运单位名称并显示出来。

程序代码如下。

```
DECLARE
  _exprname CHARACTER (128);
BEGIN
  SELECT "exprname" INTO _exprname FROM sales.orders
  WHERE ordid='149456121806';
  RAISE NOTICE '承运单位 = %', _exprname;
END;
```

程序运行结果如下。

```
NOTICE:  承运单位 = 顺丰速运
```

示例 4.2：SELECT INTO 查询语句的异常。

功能描述：本示例查询订单表中运送方式为航空的承运单位名称，并显示出来。示例中出现预定义的异常 TOO_MANY_ROWS。应确保查询返回单行，或使用 LIMIT 1 加以限制。

程序代码如下。

```
DECLARE
  _exprname CHARACTER (128);
```

```
BEGIN
  SELECT "exprname" INTO _exprname FROM sales.orders
  WHERE "mot"='航空';
  RAISE NOTICE '承运单位 = %',_exprname;
END;
```

程序运行结果如下。

```
NOTICE:   SQL 错误 [P0003]: 错误: 查询返回多条记录
```

2. SELECT BULK COLLECT INTO 语句

利用 BULK COLLECT 短语可以将数据批量地从 SQL 引擎传送给 PL/SQL 引擎,从而实现多条记录的查询,而不是每次传送一行数据。BULK COLLECT 语句减少了 PL/SQL 引擎和 SQL 引擎之间的切换次数,从而减少了提取数据时的额外开销。

BULK COLLECT 的语法格式如下。

```
SELECT... BULK COLLECT INTO collection_name[ , collection_name] ...
```

其中,collection_name 代表一个集合。使用 BULK COLLECT 语句时,有以下规则和限制。

(1) 不论动态 SQL 还是静态 SQL,都可以使用 BULK COLLECT 语句。

(2) 可以在 SELECT INTO、FETCH INTO 以及 RETURNING INTO 子句中使用。

(3) 对于在 BULK COLLECT 语句中使用的集合,SQL 引擎会自动进行初始化及扩展,从索引 1 开始填充集合,连续地插入元素,并且覆盖之前已经被使用过元素的值。若 SELECT…BULK COLLECT 没有找到任何行,不会引发 NO_DATA_FOUND 异常。所以,必须对集合的内容进行检查,确认其中是否有数据。

(4) 如果查询结果没有返回任何行,集合的 COUNT 机制将返回 0。

使用 BULK COLLECT 短语没有返回任何数据时,PL/SQL 不会抛出异常。因此,需要检查目标集合中是否有数据。

示例 4.3:通过 BULK COLLECT 获取商品 ID 及名称。

程序代码如下。

```
DECLARE
  TYPE IdTab IS TABLE OF goods.goodid%TYPE;
  TYPE NameTab IS TABLE OF goods.goodname%TYPE;

  ids IdTab;
  names NameTab;

  PROCEDURE print_first_n (n POSITIVE) IS
  BEGIN
    IF ids.COUNT = 0 THEN
      raise notice 'Collections are empty.';
    ELSE
```

```
      raise notice 'First % goods:', n;

      FOR i IN 1 .. n LOOP
        raise notice ' goods # id: %, name:%;',ids(i) ,names(i);
      END LOOP;
    END IF;
  END;

BEGIN
  SELECT goodid, goodname
  BULK COLLECT INTO ids, names
  FROM goods
  ORDER BY goodid;
  print_first_n(3);
end;
```

程序运行结果如下。

```
First 3 goods:
 goods # id: 97, name:纸飞机 diy 汽车钥匙扣;
 goods # id: 1769, name:口罩一次性;
 goods # id: 4546, name:斜跨包;
```

上述示例同样可以使用 FOR LOOP 完成相同的任务，但花费时间更长。如果商品表里有上千条记录，那么 PL/SQL 引擎就要向游标发出上千次 FETCH 操作。为了提高这种场景下的访问效率，可以在 INTO 子句中使用 BULK COLLECT 语句。在游标（显式或隐式游标）中使用这个子句，则是告知 SQL 引擎，把查询提取出来的多行数据批量绑定到指定的集合上，然后再把控制返回给 PL/SQL 引擎，能大幅减少上下文切换，从而提升程序的性能。

对 SELECT 语句，还存在另一种需求。SELECT 作为查询语句，不会产生冲突操作，一般情况下没有锁。但有时确实需要先将数据查询出来，然后再作更新，现实中也存在这种可能，即刚查询完结果还没更新，就被其他事务更新了，解决这个问题需要采用 SELECT FOR UPDATE 语句。它是标准 SELECT 的一种特殊变体，用于主动在查询提取的每一行数据上放置一个行级锁。只有当需要预先保留查询的结果时，才使用 SELECT FOR UPDATE，以确保处理数据时不会被他人更新。

默认情况下，SELECT FOR UPDATE 语句会一直等待，直到获得请求的行锁。若要更改此行为，可以在 FOR UPDATE 句中添加可选关键词 NOWAIT 或 SKIP LOCKED，告诉数据库，如果其他用户锁定了该表，则无需等待。

示例 4.4：FOR UPDATE 子句。

功能描述：下单时对该商品库存加锁。

程序代码如下。

```
BEGIN
  SELECT * FROM supply
  WHERE shopid='102108' AND goodid='69488111449'
```

```
    FOR UPDATE ;

    UPDATE supply
    SET totlwhamt=totlwhamt-1
    WHERE shopid='102108' AND goodid='69488111449';

    COMMIT;
END;
```

4.1.3　PL/SQL 中的 DML 语句

PL/SQL 中的 DML 语句对标准 SQL 语句中的 DML 语句进行扩展,允许使用变量。

当 SQL 对应部分具有一个绑定变量的占位符时,PL/SQL 静态 SQL 语句是可以具有 PL/SQL 标识符的,PL/SQL 标识符必须标识为变量或形式参数。如果要将 PL/SQL 标识符用于表名、列名等,请参考 PL/SQL 动态 SQL 的相关内容。

此外,对 PL/SQL 中的 DML 语句,还需要注意以下内容。

(1) PL/SQL 代码运行 DML 语句后,某些变量的值是未定义的。

例如,在 FETCH 或者 SELECT 语句引发异常后,该语句之后定义的变量的值未定义。在影响零行的 DML 语句之后,OUT 绑定变量的值是未定义的,除非 DML 语句是一个 BULK 或多行操作。

(2) 如果要查询当前 DML 语句操作的记录信息,可以在 DML 语句末尾使用 RETURNING 语句,返回该记录的信息。RETURNING 子句可以减少网络上的往返交互,消耗更少的服务器的 CPU 时间,并减少程序中打开和管理的游标数量。

(3) 带有 BULK COLLECT 的 RETURNING INTO 子句,也称为 RETURNING BULK COLLECT INTO 子句,用于批量操作。可以出现在 INSERT、UPDATE、DELETE 或 EXECUTE IMMEDIATE 语句中。使用 RETURNING BULK COLLECT INTO 子句,语句将其结果集存储在一个或多个集合中。

(4) 在 PL/SQL 中还可以使用记录类型作为 DML 语句的参数或变量。

示例 4.5:静态 SQL 语句。

功能描述:本例对商品表进行了 INSERT、UPDATE、DELETE 操作。将新商品添加到临时表中,并且修改商品名称,最后删除此条记录。该块还使用静态 SQL 语句 COMMIT。

程序代码如下。

```
DROP TABLE IF EXISTS goods_temp;
CREATE TABLE goods_temp AS
SELECT goodid, goodname FROM goods;

DECLARE
  g_goodid    goods.goodid%TYPE := '456610852';
  g_goodname  goods.goodname%TYPE := '牙刷';
BEGIN
```

```
    INSERT INTO goods_temp (goodid, goodname)
    VALUES (g_goodid, g_goodname);

    UPDATE goods_temp
    SET goodname = '牙刷 x6'
    WHERE goodid = g_goodid;

    DELETE FROM goods_temp
    WHERE goodid = g_goodid
    RETURNING goodname
    INTO g_goodname;

    COMMIT;
    RAISE NOTICE 'g_goodname:%',g_goodname;
END;
```

程序运行结果如下。

```
NOTICE:   g_goodname:牙刷 x6
```

将上述示例改写为采用记录类型作为 DML 语句的参数或变量，如示例 4.6 所示。

示例 4.6：使用记录类型的 DML 语句。

程序代码如下。

```
DROP TABLE IF EXISTS goods_temp;
CREATE TABLE goods_temp AS
SELECT goodid, goodname FROM goods;

DECLARE
  r_good goods_temp%ROWTYPE;
  return_good r_good%TYPE;
BEGIN
  r_good.goodid='456610852';
  r_good.goodname='牙刷';

  INSERT INTO goods_temp
  VALUES r_good;

  r_good.goodname='牙刷 x6';
  UPDATE goods_temp
  SET ROW = r_good
  WHERE goodid = r_good.goodid;

  DELETE FROM goods_temp
  WHERE goodid = r_good.goodid
  RETURNING *
  INTO return_good;

  COMMIT;
```

```
    RAISE NOTICE 'goodname:%',return_good.goodname;
END;
```

程序运行结果如下。

```
NOTICE:   goodname:牙刷 x6
```

示例 4.7：使用带有 RETURNING BULK COLLECT INTO 的 DELETE 语句，从顾客表中删除 ID 为 5000 的顾客，并在两个集合中返回它们。

程序代码如下。

```
DECLARE
  TYPE IdList IS TABLE OF customers.custid%TYPE;
  ids    IdList;
  TYPE NameList IS TABLE OF customers.custname%TYPE;
  names   NameList;
BEGIN
  DELETE FROM "customers"
  WHERE "custid" = 5000
  RETURNING "custid" , "custname"
  BULK COLLECT INTO ids, names;

  raise notice 'Deleted % rows:', SQL%ROWCOUNT;
  FOR i IN ids.FIRST .. ids.LAST
  LOOP
    raise notice 'customers # %:%', ids(i), names(i);
  END LOOP;
END;
```

程序运行结果如下。

```
Deleted 1 rows:
customers # 5000:少叔红
```

4.2 游标

4.2.1 游标概念

游标提供了一种对检索获得的数据进行操作的灵活手段，游标由结果集（可以是零条、一条或由相关的选择语句检索出的多条记录）和结果集中指向特定记录的游标位置组成。就本质而言，游标是一种从包括多条记录的结果集中每次提取一条记录进行操作的作用机制。

关系型数据库是面向集合的操作，KingbaseES 数据库中并没有一种描述表中单一记录的表达形式，除非使用 WHERE 语句限制只有一条记录被选中。因此，必须借助游标进行面向单条记录的数据处理，它提供了基于游标位置对表中数据进行删除或更新的能力。也正是游标把面向集合的数据库管理系统和面向行的程序设计结合起来,使这两种数据处理方式在应用协作方面成为可能。

通常,由 PL/SQL 构造和管理的游标称为隐式游标,用户自己构造和管理的游标称为显式游标。KingbaseES 可以从任何会话游标的属性中获取相关会话游标的有关信息。一个会话可以同时打开的游标数量由以下因素决定。

① 会话可用的内存量。

② 初始化参数的值 ORA_OPEN_CURSORS(仅 PL/SQL 内部游标生效)。

显式游标是用户构建和管理的会话游标,用于处理 SELECT 语句返回的多行数据,显示游标的使用顺序为声明、打开、关闭。即必须先声明并定义一个显式游标,为其命名,并将其与查询相关联,在执行部分或异常处理部分打开,然后通过以下任一方式处理查询结果集。

① 打开显式游标(使用 OPEN 语句),从结果集中获取行(使用 FETCH 语句),然后关闭显式游标(使用 CLOSE 语句)。

② 在游标 FOR LOOP 语句中使用显式游标。

与隐式游标不同的是,用户可以通过名称来引用显式游标或游标变量。因此,显式游标或游标变量又称为命名游标。

4.2.2 隐式游标

隐式游标是由 PL/SQL 构造和管理的会话游标,也称为 SQL 游标。

当执行一个 DML 语句,或者执行将来自数据库的数据返回到某数据结构的 SELECT INTO 语句时,PL/SQL 都会声明并管理一个隐式游标。之所以称为隐式游标,是因为数据库以隐式方式或自动处理与游标相关的许多操作,用户不能直接命令或者控制此类游标,但可以从其属性中获取与最近执行的 SQL 语句相关的信息,隐式游标属性如表 4.1 所示。

表 4.1 隐式游标属性

属性	说明
%ISOPEN	布尔值,判断游标是否打开。始终具有值 FALSE
%FOUND	布尔值,判断是否有任何行受到影响
%NOTFOUND	布尔值,判断是否没有行受到影响,与%FOUND 相反
%ROWCOUNT	数值型,判断有多少行受到影响

隐式游标属性值的语法格式如下。

```
SQL%attribute
```

SQL%attribute 总是指向最后一次执行的语句。如果没有运行这样的语句,则 SQL%attribute 的值为 NULL。

1. SQL%ISOPEN 属性：判断游标是否打开

隐式游标的 SQL%ISOPEN 永远为 FALSE，因为语句块执行完毕后，游标自动关闭。

2. SQL%FOUND 属性：判断是否有任何行受到影响

若成功影响一行或多行，则返回 TRUE；否则返回 FALSE。

3. SQL%NOTFOUND 属性：判断是否没有行受到影响

SQL%NOTFOUND，为 SQL%FOUND 的逻辑反义词，未能成功获取到一行，则返回 TRUE，否则返回 FALSE。

SQL%NOTFOUND 属性对 SELECT INTO 语句没有用处，是因为如果 SELECT INTO 语句没有返回任何行，数据库还将引发 NO_DATA_FOUND 异常。

4. SQL%ROWCOUNT 属性：判断有多少行受到影响

SQL%ROWCOUNT 用于记录受影响的函数，可以用于 SELECT INTO 或 DML 语句。当执行多条 DML 语句时，以 SQL%ROWCOUNT 之前执行的最后一条语句受影响行数为准。如果没有运行相关语句，则 SQL%ROWCOUNT 的结果为 NULL。

示例 4.8：隐式游标综合应用。

功能描述：先插入 1 条数据，然后执行两次删除给定分类编码的记录，使用 SQL%FOUND 查看是否发现删除，若成功，输出"成功删除商品类别编号"，否则输出"不存在商品类别编号"；使用 SQL%ROWCOUNT 输出删除顾客数；使用 SQL%ISOPEN 查看游标是否打开。

程序代码如下。

```
DECLARE
  new_catgid categories.catgid%TYPE;
BEGIN
  SELECT max(catgid)+1 INTO new_catgid
  FROM categories;
  --新增商品类别
  INSERT INTO categories(catgid,catgname,parentid)
  VALUES (new_catgid,'手机',1);
  --连续删除两次
  FOR i IN 1..2 LOOP
      DELETE FROM categories
      WHERE catgid=new_catgid;

      IF SQL%FOUND THEN
         RAISE NOTICE '成功删除商品类别编号 %',new_catgid;
      ELSE
         RAISE NOTICE '不存在商品类别编号 %',new_catgid;
      END IF;

   RAISE NOTICE '删除顾客数：%', SQL%ROWCOUNT;
END LOOP ;
IF SQL%ISOPEN THEN
   RAISE NOTICE '游标已打开';
```

```
    ELSE
      RAISE NOTICE '游标已关闭';
    END IF;
END
```

程序运行结果如下。

```
成功删除商品类别编号 2075
删除顾客数:1
不存在商品类别编号 2075
删除顾客数:0
游标已关闭
```

4.2.3 声明和定义显式游标

用户可以先声明一个显式游标,然后在同一个块、子程序或包中定义它,或者同时声明和定义它。

仅声明显式游标的语法格式如下。

```
CURSOR cursor_name [ parameter_list ] RETURN return_type;
```

同时声明和定义显式游标的语法格式如下。

```
CURSOR cursor_name [ parameter_list ] [ RETURN return_type ]
IS select_statement;
```

示例 4.9:显式游标声明和定义。

功能描述:声明游标 c1,然后定义女性顾客的所有信息。声明并定义游标 c2,获得男性顾客的所有信息。

程序代码如下。

```
DECLARE
  --声明 c1
  CURSOR c1 RETURN customers%ROWTYPE;
  --声明和定义 c2
  CURSOR c2 IS
    SELECT * FROM customers
    WHERE gender='男';
  --定义 c1,定义时带 return type
  CURSOR c1 RETURN customers%ROWTYPE IS
    SELECT * FROM customers
    WHERE gender='女';
BEGIN
  NULL;
END;
```

4.2.4 打开和关闭显式游标

声明和定义显式游标后,可以使用 OPEN 语句打开它,使用 CLOSE 语句关闭打开的显式游标,从而允许重用其资源。关闭游标后,将无法从其结果集中获取记录或引用其属性。如果游标关闭后尝试获取游标的信息,PL/SQL 会引发预定义的异常 INVALID_CURSOR。

已经关闭的显式游标可以重新打开。但是如果重复打开一个显式游标,PL/SQL 会引发预定义的异常 DUPLICATE_CURSOR。

4.2.5 使用显式游标获取数据

1. FETCH 语句

打开显式游标后,可以使用 FETCH 语句获取查询结果集的行。FETCH 返回一行语句的基本语法格式如下。

```
FETCH cursor_name INTO into_clause
```

into_clause 是变量列表或单个记录变量。对于查询返回的每一列,变量列表或记录必须具有对应的类型兼容变量或字段。还可以用%TYPE 和%ROWTYPE 属性更为方便地声明 FETCH 语句中使用的变量和记录。

FETCH 语句检索结果集的当前行,将该行的列值存储到变量或记录中,并将游标指向到下一行。通常,在 LOOP 语句中使用 FETCH 语句,当 FETCH 语句取完所有行时退出 LOOP 语句。FETCH 不到新的数据时,PL/SQL 不会触发异常;当 FETCH 语句返回 NULL 时,可以通过使用游标属性%NOTFOUND 检测退出条件,来判断是否将所有的数据取完。

示例 4.10:LOOP 语句中的 FETCH 语句。

功能描述:声明并定义游标 c1、c2,在 LOOP 语句中使用 FETCH 和%NOTFOUND 属性一次一行地获取两个显式游标的结果集。第 1 个 FETCH 语句将列值检索到变量中,第 2 个 FETCH 语句将列值检索到记录中。变量和记录分别使用%TYPE 和%ROWTYPE 声明。

程序代码如下。

```
DECLARE
  CURSOR c1 IS
    SELECT custid, custname FROM customers
    WHERE gender ='男' limit 3;

    v_id   customers.custid%TYPE;
    v_name customers.custname%TYPE;

  CURSOR c2 IS
    SELECT * FROM customers
    WHERE gender ='女' limit 3;
```

```
    v_customer customers%ROWTYPE;
BEGIN
  OPEN c1;
  LOOP
    FETCH c1 INTO v_id, v_name;
    EXIT WHEN c1%NOTFOUND;
    RAISE NOTICE 'id: %, name: %', v_id, v_name;
  END LOOP;
  CLOSE c1;
  RAISE NOTICE '------------------';

  OPEN c2;
  LOOP
    FETCH c2 INTO v_customer;
    EXIT WHEN c2%NOTFOUND;
    RAISE NOTICE 'id: %, name: %', v_customer.custid, v_customer.custname;
  END LOOP;
  CLOSE c2;
END;
```

程序运行结果如下。

```
Notice:id: 1, name: 林心水
Notice:id: 7, name: 石盈
Notice:id: 8, name: 梁豹
Notice:------------------
Notice:id: 2, name: 江文曜
Notice:id: 3, name: 吕浩初
Notice:id: 4, name: 萧承望
```

示例 4.11：将相同的显式游标提取到不同的变量中。

功能描述：获得 5 条顾客信息，然后使用 5 次 FETCH 语句将结果集提取到 5 个记录中，每个语句提取到 1 个不同的记录变量中。记录变量用 %ROWTYPE 声明。

程序代码如下。

```
DECLARE
  CURSOR c IS
  SELECT custid, custname, gender
  FROM customers LIMIT 5;

  cust1 c%ROWTYPE;
  cust2 c%ROWTYPE;
  cust3 c%ROWTYPE;
  cust4 c%ROWTYPE;
  cust5 c%ROWTYPE;
BEGIN
  OPEN c;
  FETCH c INTO cust1;
  FETCH c INTO cust2;
```

```
    FETCH c INTO cust3;
    FETCH c INTO cust4;
    FETCH c INTO cust5;
    CLOSE c;

    RAISE NOTICE 'cust1: %', cust1;
    RAISE NOTICE 'cust2: %', cust2;
    RAISE NOTICE 'cust3: %', cust3;
    RAISE NOTICE 'cust4: %', cust4;
    RAISE NOTICE 'cust5: %', cust5;
END;
```

程序运行结果如下。

```
Notice:cust1: (1,林心水,"男 ")
Notice:cust2: (2,江文曜,"女 ")
Notice:cust3: (3,吕浩初,"女 ")
Notice:cust4: (4,萧承望,"女 ")
Notice:cust5: (5,程同光,"女 ")
```

2. 带有 BULK COLLECT 子句的 FETCH 语句

带有 BULK COLLECT 的 FETCH 语句,也称为 FETCH BULK COLLECT 语句,它的作用是将整个结果集提取到一个或多个集合变量中。PL/SQL 为 FETCH BULK COLLECT 语句提供了一个 LIMIT 子句,它限制了从数据库提取的行的数量。语法格式如下。

```
FETCH cursor BULK COLLECT INTO ... [LIMIT rows];
```

其中,rows 可以是直接量、变量或者结果值是整数的表达式(如果为非整数,则数据库会引发 VALUE_ERROR 异常)。

对于 FETCH BULK COLLECT 语句来说,LIMIT 是非常有用的。因为这个语句可以帮助控制程序用多大内存来处理数据。比如,假设需要查询并处理 10000 行的数据,可以用 BULK COLLECT 一次性取出所有的行,然后填充到一个非常大的集合中。

示例 4.12:分别通过 3 种不同方式对 FETCH 提取出的商品信息进行输出打印。
程序代码如下。

```
CREATE type listtype as (name varchar(128), price money);

DECLARE
    TYPE NameTpe IS TABLE OF goods.goodname%TYPE;
    TYPE PriceTyp IS TABLE OF goods.price%TYPE;

    CURSOR c1 IS
        SELECT goodname, price
        FROM goods
        WHERE price::numeric > 100
```

```
      ORDER BY "goodid";

    names    NameTpe;
    prices   PriceTyp;

    TYPE Typ IS TABLE OF listtype;
    recs Typ;
    v_limit PLS_INTEGER := 6;

    PROCEDURE print_results IS
    BEGIN
      --Check if collections are empty:
      IF names IS NULL OR names.COUNT = 0 THEN
        raise notice 'No results!';
          ELSE
        raise notice 'Result: ';
        FOR i IN names.FIRST .. names.LAST
        LOOP
          raise notice '%##goods#name: %, price: %.', i,names(i), prices(i);
        END LOOP;
      END IF;
    END;
BEGIN
  raise notice '--- Processing all results simultaneously ---';
  OPEN c1;
  FETCH c1 BULK COLLECT INTO names, prices;
  CLOSE c1;
  print_results();

  raise notice '--- Processing % rows at a time ---', v_limit;
  OPEN c1;
  LOOP
    FETCH c1 BULK COLLECT INTO names, prices LIMIT v_limit;
    EXIT WHEN names.COUNT = 0;
    print_results();
  END LOOP;
  CLOSE c1;

  raise notice '--- Fetching records rather than columns ---';
  OPEN c1;
  FETCH c1 BULK COLLECT INTO recs;
  FOR i IN recs.FIRST .. recs.LAST
  LOOP
    --Now all columns from result set come from one record
    raise notice '   goods#name#price %:%', recs(i).name, recs(i).price;
  END LOOP;
END;

drop type listtype;
```

程序运行结果如下。

```
--- Processing all results simultaneously ---
Result:
1#goods#name: 红蜻蜓皮鞋, price: $239.00.
2#goods#name: 知识图谱:方法、实践与应用(博文视点出品), price: $116.80.
3#goods#name: ATS 网球拍单人初学者大学生专业碳素双打男女士正品全回弹训练器, price:
$159.00.
4#goods#name: 牧予品牌女士休闲运动套装女装, price: $169.00.
5#goods#name: 小迷糊护肤套装礼盒装, price: $119.90.
6#goods#name: 玖慕 SZ002, price: $128.00.
7#goods#name: Keep B3 手环, price: $269.00.
8#goods#name: 朱顶红蜡球大球, price: $148.00.
……
```

上例通过 FETCH cursor BULK COLLECT INTO 子句将价格大于 100 的所有商品进行了 3 种方法的输出打印。第 1 种方式将所有商品的商品名和价格放入对应的两个集合中去,再通过遍历集合的方式展示所有信息;方式 2 与方式 1 的区别是将所有符合条件的商品信息每次提取 6 个放入集合,再进行输出打印,这种方式很好地避免了内存溢出等问题;方式 3 则是将所有符合条件的商品数据放入一个自定义数据集合,看起来更加简洁。

需要补充的是,当 SELECT FOR UPDATE 与显式游标关联时,该游标称为 FOR UPDATE 游标。只有 FOR UPDATE 游标可以出现在 UPDATE 或 DELETE 语句的 CURRENT OF 子句中(CURRENT OF 子句是 SQL 语句 UPDATE 和 DELETE 的 WHERE 子句的 PL/SQL 扩展,将语句限制为游标的当前行)。

示例 4.13:FOR UPDATE 子句。

功能描述:店铺商品降价处理,价格超过 60 的,调整为 50;不超过 60 的,打八折。

程序代码如下。

```
DECLARE
  s_shopid supply.shopid%TYPE:='197133';
  CURSOR c1 IS
  SELECT * FROM supply WHERE shopid=s_shopid
    FOR UPDATE NOWAIT ;
  s_price supply.price%TYPE;
BEGIN
  FOR s_supply IN c1
  LOOP
    IF s_supply.price>('60'::money) THEN
      s_price=50;
    ELSE
      s_price=s_supply.price * (0.8);
    END IF;

    UPDATE supply SET price=s_price
    WHERE CURRENT OF c1;
  END LOOP;
END;
```

4.2.6 显式游标查询中的变量

显式游标查询可以引用其范围内的任何变量。当用户打开一个显式游标时，PL/SQL 识别范围内的任何变量，并在结果集中使用这些值，稍后更改这些变量的值，并不影响结果集。只有关闭显式游标，重新打开显式游标时，才会根据修改后的变量的值生成新的查询结果集。

示例 4.14：显式游标查询中的变量——无结果集更改。

功能描述：获得店铺供应商品的库存总数量，显式游标查询引用了变量因子。当游标打开时，num 的值为 100。因此，amount_add_num 始终比 amount 多 100，尽管该 num 在每次循环后都会递增。

程序代码如下。

```
DECLARE
  amount      supply.totlwhamt%TYPE;
  amount_add_num supply.totlwhamt%TYPE;
  num         INTEGER := 100;

CURSOR c1 IS
    SELECT totlwhamt,totlwhamt+num
    FROM supply LIMIT 5;
BEGIN
  OPEN c1;
  LOOP
    FETCH c1 INTO amount, amount_add_num;
    EXIT WHEN c1%NOTFOUND;
    RAISE NOTICE 'num = %', num;
    RAISE NOTICE 'amount = %', amount;
    RAISE NOTICE 'amount_add_num = %', amount_add_num;
    RAISE NOTICE '------------------';
    --不影响 amount_add_num 的值
    num := num + 100;
  END LOOP;
  CLOSE c1;
END;
```

程序运行结果如下。

```
Notice:num = 100
Notice:amount = 1000
Notice:amount_add_num = 1100
Notice:------------------
Notice:num = 200
Notice:amount = 999
Notice:amount_add_num = 1099
Notice:------------------
Notice:num = 300
Notice:amount = 800
```

```
Notice:amount_add_num = 900
Notice:------------------
Notice:num = 400
Notice:amount = 1111
Notice:amount_add_num = 1211
Notice:------------------
Notice:num = 500
Notice:amount = 1231
Notice:amount_add_num = 1331
Notice:------------------
```

上述示例中,如果要根据 num 的值更改结果集,必须先关闭游标,更改变量的值,然后再次打开游标,如以下示例。

示例 4.15:显式游标查询中的变量——结果集更改。

功能描述:获得店铺供应商品的库存总数量,使每个商品的库存数量增加 100。

程序代码如下。

```
DECLARE
  amount      supply.totlwhamt%TYPE;
  amount_add_num supply.totlwhamt%TYPE;
  num         INTEGER := 100;

CURSOR c1 IS
  SELECT totlwhamt,totlwhamt+num
  FROM supply LIMIT 5;
BEGIN
  OPEN c1;
  LOOP
    FETCH c1 INTO amount, amount_add_num;
    EXIT WHEN c1%NOTFOUND;
    RAISE NOTICE 'amount = %', amount;
    RAISE NOTICE 'amount_add_num = %', amount_add_num;
  END LOOP;
  CLOSE c1;

  RAISE NOTICE '------------------';
  num := num + 100;
  RAISE NOTICE 'num = %', num;
  OPEN c1;
  LOOP
    FETCH c1 INTO amount, amount_add_num;
    EXIT WHEN c1%NOTFOUND;
    RAISE NOTICE 'amount = %', amount;
    RAISE NOTICE 'amount_add_num = %', amount_add_num;
  END LOOP;
  CLOSE c1;
END;
```

程序运行结果如下。

```
Notice:amount = 1000
Notice:amount_add_num = 1100
Notice:amount = 999
Notice:amount_add_num = 1099
Notice:amount = 800
Notice:amount_add_num = 900
Notice:amount = 1111
Notice:amount_add_num = 1211
Notice:amount = 1231
Notice:amount_add_num = 1331
Notice:-------------------
Notice:num = 200
Notice:amount = 1000
Notice:amount_add_num = 1200
Notice:amount = 999
Notice:amount_add_num = 1199
Notice:amount = 800
Notice:amount_add_num = 1000
Notice:amount = 1111
Notice:amount_add_num = 1311
Notice:amount = 1231
Notice:amount_add_num = 1431
```

4.2.7 当显式游标查询需要列别名时

当显式游标查询包含虚拟列(表达式)时,若满足以下任一情况,则该列须具有别名:
① 需要使用游标来获取使用 %ROWTYPE 声明的记录。
② 需要在程序中引用虚拟列。

示例 4.16:显式游标中的虚拟列需要一个别名。

功能描述:将供应表中物品的原价格(price)与折扣价格(discount)相减,取别名为最终价格(final_price)。

程序代码如下。

```
DECLARE
  CURSOR c1 IS
    SELECT goodid,shopid,price-discount::money final_price
    FROM supply
    LIMIT 3;
  supply_rec c1%ROWTYPE;
BEGIN
  OPEN c1;
  LOOP
    FETCH c1 INTO supply_rec;
    EXIT WHEN c1%NOTFOUND;
      RAISE NOTICE 'goodid:% shopid:% final_price:%',
      supply_rec.goodid, supply_rec.shopid, supply_rec.final_price;
  END LOOP;
```

```
    CLOSE c1;
END;
```

程序运行结果如下。

```
NOTICE:goodid:12560557     shopid:104142   final_price:$115.81
NOTICE:goodid:68327953786  shopid:781872   final_price:$48.81
NOTICE:goodid:100232901206 shopid:781872   final_price:$97.01
```

4.2.8 接收参数的显式游标

用户也可以定义一个具有形参的显式游标，然后在每次打开游标时将不同的实际参数传递给游标。在游标内部，可以在任何使用常量的地方使用该形参。在游标查询之外，不能引用该形参。

示例 4.17：显式游标的参数处理。

功能描述：创建了 1 个显式游标，其形式参数代表它的评价等级。查询相应评价等级记录集。然后，创建 1 个打印游标查询结果集的过程。最后，传入实参打开游标，打印结果集，关闭游标，用不同的实参打开游标，打印结果集，关闭游标。

程序代码如下。

```
DECLARE
  CURSOR c (_feeling comments.feeling%TYPE) IS
    SELECT goodid,custid,remark
    FROM comments
    WHERE feeling=_feeling LIMIT 3;

  PROCEDURE show_badfeeling IS
    _goodid comments.goodid%TYPE;
    _custid comments.custid%TYPE;
    _remark comments.remark%TYPE;
  BEGIN
    LOOP
      FETCH c INTO _goodid, _custid, _remark;
      EXIT WHEN c%NOTFOUND;
        RAISE NOTICE 'goodid: %, custid: %, remark: %',
          _goodid, _custid, _remark;
    END LOOP;
  END show_badfeeling;

  feeling_level comments.feeling%TYPE;
BEGIN
  feeling_level := '中评';
  RAISE NOTICE 'feeling_level = %',feeling_level;
  OPEN c(feeling_level);
  show_badfeeling;
  CLOSE c;
```

```
    RAISE NOTICE '------------------------------------------';
    feeling_level := '差评';
    RAISE NOTICE 'feeling_level = %',feeling_level;
    OPEN c(feeling_level);
    show_badfeeling;
    CLOSE c;
END;
```

程序运行结果如下。

```
NOTICE:feeling_level = 中评
NOTICE:goodid: 69488111449 , custid: 3, remark: 还好吧 ,也没什么作用
NOTICE:goodid: 100007325720, custid: 5, remark: 面膜用完会有红痒的情况
NOTICE:goodid: 100012178380, custid: 5, remark: 物流很快 一天就能到 包装很好看
NOTICE: ---------------------------------------
NOTICE:feeling_level = 差评
NOTICE:goodid: 667722400852, custid: 7, remark: 日期快到期了,还发货,太无语了
NOTICE:goodid: 521933550230, custid: 18, remark: 声音一个不是很脆
NOTICE:goodid: 521933550231, custid: 18, remark: 声音一个不是很脆
```

1. 具有默认值的形参游标参数

当创建带形参的显式游标时,可以为形参指定默认值。当形参有默认值时,其对应的实参是可选的。如果打开游标而不指定实参,则形参使用默认值。

示例 4.18：具有默认值的游标参数。

功能描述：查询购物评价等级为默认值"差评",以及给定值为"中评"的相关评价。

程序代码如下。

```
DECLARE
  CURSOR c (_feeling comments.feeling%TYPE DEFAULT '差评') IS
    SELECT goodid,custid,remark
    FROM comments
    WHERE feeling=_feeling LIMIT 3;

  PROCEDURE show_badfeeling IS
    _goodid comments.goodid%TYPE;
    _custid comments.custid%TYPE;
    _remark comments.remark%TYPE;
  BEGIN
    LOOP
      FETCH c INTO _goodid, _custid, _remark;
      EXIT WHEN c%NOTFOUND;
      RAISE NOTICE 'goodid: %, custid: %, remark: %',
       _goodid ,_custid ,_remark;
    END LOOP;
  END show_badfeeling;

  feeling_level comments.feeling%TYPE;
BEGIN
```

```
  RAISE NOTICE 'default feeling_level';
  OPEN c();
  show_badfeeling;
  CLOSE c;

  RAISE NOTICE '------------------------------------------';
  feeling_level := '中评';
  RAISE NOTICE 'feeling_level = %',feeling_level;
  OPEN c(feeling_level);
  show_badfeeling;
  CLOSE c;
END;
```

程序运行结果如下。

```
NOTICE:default feeling_level
NOTICE:goodid: 667722400852, custid: 7, remark: 日期快到期了,还发货,太无语了
NOTICE:goodid: 521933550230, custid: 18, remark: 声音一个不是很脆
NOTICE:goodid: 521933550231, custid: 18, remark: 声音一个不是很脆
NOTICE:------------------------------------------
NOTICE:feeling_level = 中评
NOTICE:goodid: 69488111449, custid: 3, remark: 还好吧 ,也没什么作用
NOTICE:goodid: 100007325720, custid: 5, remark: 面膜用完会有红痒的情况
NOTICE:goodid: 100012178380, custid: 5, remark: 物流很快 一天就能到 包装很好看
```

2. 添加带默认值的形参游标参数

如果为游标添加一项形参,并且为新增的参数指定默认值,则之前对游标引用不需要作任何更改。请比较示例 4.19 和 4.18 的不同之处。

示例 4.19:向现有游标添加形式参数。

功能描述:查询购物评价等级为给定值,并且评价提交时间大于给定时间的评价记录。

程序代码如下。

```
DECLARE
  CURSOR c (_feeling bpchar(4),_submitdate timestamp DEFAULT to_date('2021-01-
01','yyyy-MM-dd') ) IS
  SELECT goodid,custid,remark
  FROM comments
  WHERE feeling=_feeling and submtime>_submitdate LIMIT 3;

  PROCEDURE show_badfeeling IS
    _goodid comments.goodid%TYPE;
    _custid comments.custid%TYPE;
    _remark comments.remark%TYPE;
  BEGIN
    LOOP
      FETCH c INTO _goodid, _custid, _remark;
      EXIT WHEN c%NOTFOUND;
      RAISE NOTICE 'goodid: %, custid: %, remark: %',
```

```
        _goodid , _custid , _remark;
END LOOP;
  END show_badfeeling;

  feeling_level comments.feeling%TYPE;
  submit_date comments.submtime%TYPE;
BEGIN
  feeling_level := '差评';
  OPEN c(feeling_level);
  show_badfeeling;
  CLOSE c;

  RAISE NOTICE '-------------------------------------------';
  feeling_level := '中评';
  submit_date:=to_date('2022-01-01','yyyy-MM-dd');
  RAISE NOTICE 'feeling_level = %',feeling_level;
  OPEN c(feeling_level,submit_date);
  show_badfeeling;
  CLOSE c;
END;
```

程序运行结果如下。

```
NOTICE:goodid: 667722400852, custid: 7, remark: 日期快到期了,还发货,太无语了
NOTICE:goodid: 521933550230, custid: 18, remark: 声音一个不是很脆
NOTICE:goodid: 521933550231, custid: 18, remark: 声音一个不是很脆
NOTICE:-----------------------------------------
NOTICE:feeling_level = 中评
NOTICE:goodid: 660223524680, custid: 7, remark: 非常好,除了其中两盒快到期外,其他到
期日都是很后,包装好
NOTICE:goodid: 2015158083  , custid: 2593, remark: 有无异味:无异味
NOTICE:goodid: 9008        , custid: 4354, remark: 有无异味:没异味。
```

4.2.9 显式游标属性

显式游标的属性如表 4.2 所示。

表 4.2 显式游标属性

属性	说明
%ISOPEN	布尔值,判断游标是否打开
%FOUND	布尔值,判断是否有任何行受到影响
%NOTFOUND	布尔值,判断是否没有行受到影响,与%FOUND 相反
%ROWCOUNT	数值型,判断有多少行受到影响

1. %ISOPEN 属性:判断游标是否打开

该属性检测游标是否打开,如果已经打开,则返回 TRUE。%ISOPEN 属性适用于如下

情况。

① 在尝试打开显式游标之前检查它是否尚未打开。

② 如果试图打开一个已经打开的显式游标，PL/SQL 会引发预定义的异常 CURSOR_ALREADY_OPEN。必须先关闭显式游标，然后才能重新打开它。

③ 在尝试关闭显式游标之前检查它是否打开。

示例 4.20：%ISOPEN 显式游标属性。

功能描述：从商品分类表中获得记录集，作为游标 c1，仅当游标 c1 未打开时才打开它，仅当打开时才将其关闭。

程序代码如下。

```
DECLARE
  CURSOR c1 IS
    SELECT catgid, catgname
    FROM categories LIMIT 3;

  id    categories.catgid%TYPE;
  name  categories.catgname%TYPE;
BEGIN
  IF NOT c1%ISOPEN THEN
    OPEN c1;
  END IF;

  FETCH c1 INTO id, name;

  IF c1%ISOPEN THEN
    CLOSE c1;
  END IF;
END;
```

2. %FOUND 属性：判断游标是否读取到数据

%FOUND 属性表示当前游标是否指向有效的一行。在显式游标打开之后，但还未提取数据之前返回 NULL，如果最近从显式游标提取数据返回一行，则为 TRUE，否则为 FALSE，该属性一般用于判断是否结束当前游标的使用。

示例 4.21：%FOUND 显式游标属性。

功能描述：循环遍历 1 个结果集（商品分类记录集），如果找到，打印每个提取的行，并在没有更多行可提取时关闭游标。

程序代码如下。

```
DECLARE
  CURSOR c1 IS
    SELECT catgid, catgname
    FROM categories LIMIT 5;

  id    categories.catgid%TYPE;
  name  categories.catgname%TYPE;
```

```
BEGIN
  OPEN c1;
  LOOP
    FETCH c1 INTO id, name;
    IF c1%FOUND THEN
      RAISE NOTICE 'id = %, name = %',id,name;
    ELSE
      EXIT;
    END IF;
  END LOOP;
END;
```

3. %NOTFOUND 属性：判断游标是否未读取到数据

%NOTFOUND 属性(与%FOUND 属性逻辑相反)表示当前游标是否未指向有效的一行。在显式游标打开之后，但还未提取数据之前返回 NULL，如果最近从显式游标提取数据返回一行，则为 FALSE，否则为 TRUE，该属性对于退出循环的条件判断很有用。请读者比较示例 4.22 和 4.21 的不同之处。

示例 4.22：%NOTFOUND 显式游标属性。

功能描述：循环遍历一个结果集(商品分类记录集)，如果没有找到，退出并进入下一次循环，直到所有记录遍历结束。

程序代码如下。

```
DECLARE
  CURSOR c1 IS
    SELECT catgid,catgname
    FROM categories LIMIT 5;

  id   categories.catgid%TYPE;
  name categories.catgname%TYPE;
BEGIN
  OPEN c1;
  LOOP
    FETCH c1 INTO id, name;
    IF c1%NOTFOUND THEN
      EXIT;
    ELSE
      RAISE NOTICE 'id = %, name = %',id,name;
    END IF;
  END LOOP;
END;
```

4. %ROWCOUNT 判断有多少行受到影响

%ROWCOUNT 返回受影响的行数，在显式游标打开之后，但在第一次获取数据之前为 0，否则将返回获取的行数。

示例 4.23：%ROWCOUNT 显式游标属性。

功能描述：对商品分类记录集的行进行编号,并打印,并在获取第 3 行后打印 1 条消息。

程序代码如下。

```
DECLARE
  CURSOR c1 IS
    SELECT catgid,catgname FROM categories LIMIT 5;

  id   categories.catgid%TYPE;
  name categories.catgname%TYPE;
BEGIN
  OPEN c1;
  LOOP
    FETCH c1 INTO id, name;
    EXIT WHEN c1%NOTFOUND;
    RAISE NOTICE '%. %',c1%ROWCOUNT ,name;
    if c1%ROWCOUNT = 3 THEN
       RAISE NOTICE 'FETCH 3 ROWS';
    END IF;
  END LOOP;
END;
```

程序运行结果如下。

```
NOTICE:1.数字杂志
NOTICE:2.多媒体图书
NOTICE:3.音像
NOTICE:FETCH 3 ROWS
NOTICE:4.音乐
NOTICE:5.影视
```

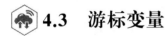 4.3 游标变量

显式游标在定义时与特定的查询相关联,其结构是不变的,因此又称为静态游标。游标变量是一个指向查询结果集的指针,不与特定的查询关联,因此具有更大的灵活性。

使用游标变量,可以完成以下事项。

① 在程序执行中,将其与不同时间的不同查询相联系。换句话说,一个游标变量可以用于从不同结果集获取。

② 将它们作为参数传递给程序或函数。在本质上,可以通过传递引用给结果集来共享该游标结果。

③ 使用静态 PL/SQL 游标的完整功能。可以在 PL/SQL 程序中使用游标变量进行 OPEN、CLOSE 和 FETCH 操作。也可以为游标变量引用标准游标属性——%ISOPEN、%FOUND、%NOTFOUND 和%ROWCOUNT。

④ 将游标内容(及其结果集)赋值给其他游标变量。游标变量是一个变量,因此它可以用于赋值操作。

游标变量和游标一样,也是一个指向多行结果集的指针,但是游标与游标变量是不同的,就像常量与变量的关系一样。它们的特性如下。

① 游标只能与指定的查询关联,即固定指向一个查询的内存处理区域,而游标变量则可与不同的查询语句关联。它可以指向不同查询语句的内存处理区域(但不能同时指向多个内存处理区域,某一时刻只能与一个查询语句相连),只要这些查询语句的返回类型兼容即可。

② 游标变量可以在表达式中使用。

③ 游标变量可以是子程序参数,也可以使用游标变量在子程序之间传递查询结果集。

④ 可以是主机变量,可以使用游标变量在 PL/SQL 存储子程序及其客户端之间传递查询结果集。

⑤ 游标变量不能接受参数。不能将参数传递给游标变量,但可以将整个查询传递给它。查询语句可以包括变量。

在引用游标变量之前,必须使其指向一个 SQL 工作区,可以通过打开它或为其分配一个打开的 PL/SQL 游标变量,或打开的宿主游标变量值的方式。

4.3.1 创建游标变量

要创建游标变量,有如下两种方式:
① 使用预定义游标引用类型 SYS_REFCURSOR。
② 先定义 REF CURSOR 类型,然后再声明该类型的变量。
REF CURSOR 类型定义的基本语法格式如下。

```
TYPE type_name IS REF CURSOR [ RETURN return_type ]
```

- 如果指定了 return_type,则该 REF CURSOR 类型和游标变量称为强游标。
- 如果没有指定 return_type,则称为弱游标,SYS_REFCURSOR 和该类型的游标变量是弱游标。

强游标的优势在于,编译器可以确定开发人员是否已经正确匹配了游标变量 FETCH 语句和其游标对象查询列表。

弱游标变量比强游标变量更易于出错,但更为灵活。使用强游标变量,只能关联返回指定类型的查询。使用弱游标变量,可以关联任何查询。弱 REF CURSOR 类型可以相互交换,也可以与预定义类型 SYS_REFCURSOR 互换。可以将弱游标变量的值分配给任何其他弱游标变量。

只有当两个游标变量具有相同的类型(不仅仅是相同的返回类型)时,才能将强游标变量的值分配给另一个强游标变量。

示例 4.24:游标变量声明。

功能描述:本例分别定义了强和弱 REF CURSOR 类型、各自的变量以及预定义类型 SYS_REFCURSOR 的变量。

程序代码如下。

```
DECLARE
  TYPE strong_cursor IS REF CURSOR RETURN customers%ROWTYPE;
  TYPE weak_cursor IS REF CURSOR;
  cursor1 strong_cursor;              --强游标
  cursor2 weak_cursor;                --弱游标
  my_cursor SYS_REFCURSOR;            --弱游标
BEGIN
  NULL;
END;
```

示例 4.25：具有用户定义的返回类型的游标变量。

功能描述：在本例中，CustRecType 是用户定义的 RECORD 类型。

程序代码如下。

```
DECLARE
  TYPE CustRecType IS RECORD (
    id NUMBER,
    name TEXT);

  TYPE CustCurType IS REF CURSOR RETURN CustRecType;
  cust_cv CustCurType;
BEGIN
  NULL;
END;
```

4.3.2 打开和关闭游标变量

声明游标变量后，可以使用 OPEN FOR 语句打开它，该语句执行以下操作。

（1）将游标变量与查询相关联（通常，查询返回多行），查询可以包含绑定变量的占位符，在 OPEN FOR 语句的 USING 子句中指定其值。

（2）分配数据库资源来处理查询。

（3）处理查询，其中包括：

① 标识结果集，如果查询引用变量，它们的值会影响结果集。

② 如果查询有 FOR UPDATE 子句，则锁定结果集的行。

（4）将游标定位在结果集的第一行之前。

在重新打开游标变量之前，不需要关闭它，即在另一个 OPEN FOR 语句中使用它。重新打开游标变量后，之前与其关联的查询将丢失。

当不再需要游标变量时，可以使用 CLOSE 语句将其关闭，从而允许重用其资源。关闭游标变量后，将无法再从其结果集中获取记录或引用其属性。否则 PL/SQL 会引发预定义异常 INVALID_CURSOR，除非重新打开已关闭的游标变量。

4.3.3 使用游标变量获取数据

打开游标变量后，可以使用 FETCH 语句获取查询结果集的行。

游标变量的返回类型必须与 FETCH 语句的 into_clause 兼容。

① 如果游标变量为强游标,PL/SQL 会在编译时捕获不兼容性。

② 如果游标变量是弱游标,PL/SQL 会在运行时捕获不兼容性,在第一次 FETCH 之前引发预定义的异常 ROWTYPE_MISMATCH。

示例 4.26:使用游标变量获取数据。

功能描述:从顾客信息表中获取记录集保存到游标变量中,并通过游标变量输出数据。

程序代码如下。

```
DECLARE
  cv SYS_REFCURSOR;
  c_id    customers.custid%TYPE;
  c_name customers.custname%TYPE;
  query TEXT := 'SELECT * FROM customers LIMIT 5';
  v_cust customers%ROWTYPE;

BEGIN
  OPEN cv FOR
    SELECT custid, custname FROM customers
    LIMIT 3;

  LOOP
    FETCH cv INTO c_id, c_name;
    EXIT WHEN cv%NOTFOUND;
    RAISE NOTICE 'c_id: %, c_name: %', c_id, c_name;
  END LOOP;

  RAISE NOTICE '----------------------------------------';

  OPEN cv FOR query;
  LOOP
    FETCH cv INTO v_cust;
    EXIT WHEN cv%NOTFOUND;
      RAISE NOTICE 'c_id: %, c_name: %', v_cust.custid, v_cust.custname;
  END LOOP;

  CLOSE cv;
END;
```

程序运行结果如下。

```
NOTICE:c_id: 1, c_name: 林心水
NOTICE:c_id: 2, c_name: 江文曜
NOTICE:c_id: 3, c_name: 吕浩初
NOTICE:------------------------------------
NOTICE:c_id: 1, c_name: 林心水
NOTICE:c_id: 2, c_name: 江文曜
NOTICE:c_id: 3, c_name: 吕浩初
NOTICE:c_id: 4, c_name: 萧承望
NOTICE:c_id: 5, c_name: 程同光
```

示例 4.27：从游标变量提取到集合中。

功能描述：使用 FETCH 语句的 BULK COLLECT 子句，从游标变量中提取到 2 个集合（嵌套表）中。

程序代码如下。

```
DECLARE
  TYPE custcurtype IS REF CURSOR;
  TYPE custnamelist IS TABLE OF customers.custname%TYPE;
  TYPE idlist IS TABLE OF customers.custid%TYPE;
  cust_cv    custcurtype;
  custnames custnamelist;
  custids    idlist;
BEGIN
  OPEN cust_cv FOR
    SELECT custname, custid FROM customers
    LIMIT 5;

  FETCH cust_cv BULK COLLECT INTO custnames, custids;
  CLOSE cust_cv;

  FOR i IN custnames.FIRST .. custnames.LAST
  LOOP
    RAISE NOTICE 'custname = %, custid = %', custnames(i), custids(i);
  END LOOP;
END;
```

程序运行结果如下。

```
NOTICE:custname = 林心水, custid = 1
NOTICE:custname = 江文曜, custid = 2
NOTICE:custname = 吕浩初, custid = 3
NOTICE:custname = 萧承望, custid = 4
NOTICE:custname = 程同光, custid = 5
```

4.3.4 为游标变量赋值

可以将另一个 PL/SQL 游标变量或宿主游标变量的值赋给 PL/SQL 游标变量。语法格式如下。

```
target_cursor_variable := source_cursor_variable;
```

- 如果 source_cursor_variable 是打开状态，则在赋值后，target_cursor_variable 也是打开的。两个游标变量指向同一个 SQL 工作区。
- 如果 source_cursor_variable 没有打开，赋值后打开 target_cursor_variable，并不会打开 source_cursor_variable。

4.3.5 游标变量查询中的变量

与游标变量关联的查询可以引用其范围内的任何变量。

当使用 OPEN FOR 语句打开游标变量时,PL/SQL 会计算查询中的任何变量,并在识别结果集时使用这些值。稍后更改变量的值不会更改结果集。如示例 4.28 所示。

要更改结果集,必须更改变量的值,然后为相同的查询再次打开游标变量,如示例 4.29 所示。请注意二者的差别。

示例 4.28:游标变量查询中的变量——无结果集更改。

功能描述:当游标变量打开时,店铺商品供应表中的 num 值为 100。因此,amount_add_num 始终比 amount 多 100,尽管该 num 在每次循环后都会递增。

程序代码如下。

```
DECLARE
  amount      supply.totlwhamt%TYPE;
  amount_add_num supply.totlwhamt%TYPE;
  num         INTEGER := 100;
  cv SYS_REFCURSOR;
BEGIN
  OPEN cv FOR
    SELECT totlwhamt,totlwhamt+num
    FROM supply LIMIT 5;

  LOOP
    FETCH cv INTO amount, amount_add_num;
    EXIT WHEN cv%NOTFOUND;
    RAISE NOTICE 'num = %', num;
    RAISE NOTICE 'amount = %', amount;
    RAISE NOTICE 'amount_add_num = %', amount_add_num;
    RAISE NOTICE '-------------------';
    --不影响 amount_add_num 的值
    num := num + 100;
  END LOOP;
  CLOSE cv;
END;
```

程序运行结果如下。

```
NOTICE:num = 100
NOTICE:amount = 1000
NOTICE:amount_add_num = 1100
NOTICE:-------------------
NOTICE:num = 200
NOTICE:amount = 999
NOTICE:amount_add_num = 1099
NOTICE:-------------------
NOTICE:num = 300
NOTICE:amount = 800
```

```
NOTICE:amount_add_num = 900
NOTICE:------------------
NOTICE:num = 400
NOTICE:amount = 1111
NOTICE:amount_add_num = 1211
NOTICE:------------------
NOTICE:num = 500
NOTICE:amount = 1231
NOTICE:amount_add_num = 1331
NOTICE:------------------
```

示例 4.29：游标变量查询中的变量——结果集更改。

程序代码如下。

```
DECLARE
  amount      supply.totlwhamt%TYPE;
  amount_add_num supply.totlwhamt%TYPE;
  num         INTEGER := 100;
  cv SYS_REFCURSOR;
BEGIN
  OPEN cv FOR
  SELECT totlwhamt,totlwhamt+num
  FROM supply LIMIT 5;
  LOOP
    FETCH cv INTO amount, amount_add_num;
    EXIT WHEN cv%NOTFOUND;
    RAISE NOTICE 'amount = %', amount;
    RAISE NOTICE 'amount_add_num = %', amount_add_num;
  END LOOP;

  RAISE NOTICE '------------------';
  num := num + 100;
  RAISE NOTICE 'num = %', num;
  OPEN cv FOR
  SELECT totlwhamt,totlwhamt+num
  FROM supply LIMIT 5;
  LOOP
    FETCH cv INTO amount, amount_add_num;
    EXIT WHEN cv%NOTFOUND;
    RAISE NOTICE 'amount = %', amount;
    RAISE NOTICE 'amount_add_num = %', amount_add_num;
  END LOOP;
  CLOSE cv;
END;
```

程序运行结果如下。

```
NOTICE:amount = 1000
NOTICE:amount_add_num = 1100
```

```
NOTICE:amount = 999
NOTICE:amount_add_num = 1099
NOTICE:amount = 800
NOTICE:amount_add_num = 900
NOTICE:amount = 1111
NOTICE:amount_add_num = 1211
NOTICE:amount = 1231
NOTICE:amount_add_num = 1331
NOTICE:------------------
NOTICE:num = 200
NOTICE:amount = 1000
NOTICE:amount_add_num = 1200
NOTICE:amount = 999
NOTICE:amount_add_num = 1199
NOTICE:amount = 800
NOTICE:amount_add_num = 1000
NOTICE:amount = 1111
NOTICE:amount_add_num = 1311
NOTICE:amount = 1231
NOTICE:amount_add_num = 1431
```

4.3.6 游标变量属性

游标变量具有与显式游标相同的属性。游标变量属性值的语法是 cursor_variable_name 后跟属性(例如：cv%ISOPEN)。如果游标变量未打开，则引用除 %ISOPEN 之外的任何属性都会引发预定义异常 INVALID_CURSOR。

4.3.7 游标变量作为子程序参数

游标变量可以作为子程序参数，这对于子程序间传递查询结果很有用。例如：
- 可以在一个子程序中打开一个游标变量，并在另一个子程序中处理它。
- 在多语言应用程序中，PL/SQL 子程序可以使用游标变量将结果集返回给以不同语言编写的子程序。

将游标变量声明为子程序的形参时，如果子程序打开或为游标变量赋值，则参数模式必须为 IN OUT；如果子程序只取回或关闭游标变量，则参数模式可以是 IN 或 IN OUT。

相应的形式和实际游标变量参数必须具有兼容的返回类型。否则，PL/SQL 会引发预定义异常 ROWTYPE_MISMATCH。

在不同 PL/SQL 单元的子程序之间传递游标变量参数，请在包中定义参数的 REF CURSOR 类型。当类型在一个包中时，多个子程序可以使用它。一个子程序可以声明该类型的形式参数，而其他子程序可以声明该类型的变量，并将它们传递给第一个子程序。

示例 4.30：查询打开游标变量的过程。

功能描述：包中定义了 REF CURSOR 类型和打开该类型游标变量参数的过程。

程序代码如下。

```
CREATE OR REPLACE PACKAGE cust_data AUTHID DEFINER AS
```

```
    TYPE custcurtype IS REF CURSOR RETURN customers%ROWTYPE;
    PROCEDURE open_cust_cv (cust_cv IN OUT custcurtype);
END cust_data;

CREATE OR REPLACE PACKAGE BODY cust_data AS
    PROCEDURE open_cust_cv (cust_cv IN OUT custcurtype) IS
    BEGIN
        OPEN cust_cv FOR SELECT * FROM customers;
    END open_cust_cv;
END cust_data;
```

示例 4.31：查询打开游标变量（返回类型相同）。

功能描述：在本例中，存储过程为所选查询打开的游标变量参数，查询具有相同的返回类型。

程序代码如下。

```
CREATE OR REPLACE PACKAGE comm_data AUTHID DEFINER AS
    TYPE commcurtype IS REF CURSOR RETURN comments%ROWTYPE;
    PROCEDURE open_comm_cv (comm_cv IN OUT commcurtype, choice INT);
END comm_data;

CREATE OR REPLACE PACKAGE BODY comm_data AS
    PROCEDURE open_comm_cv (comm_cv IN OUT commcurtype, choice INT) IS
    BEGIN
        IF choice = 1 THEN
            OPEN comm_cv FOR SELECT *
            FROM comments
            WHERE feeling ='好评';
        ELSIF choice = 2 THEN
            OPEN comm_cv FOR SELECT *
            FROM comments
            WHERE feeling ='中评';
        ELSIF choice = 3 THEN
            OPEN comm_cv FOR SELECT *
            FROM comments
            WHERE feeling ='差评';
        END IF;
    END;
END comm_data;
```

示例 4.32：查询打开游标变量（不同的返回类型）。

功能描述：在本例中，存储过程为所选查询打开的游标变量参数，查询具有不同的返回类型。

程序代码如下。

```
CREATE OR REPLACE PACKAGE comm_data AUTHID DEFINER AS
    TYPE commcurtype IS REF CURSOR;
    PROCEDURE open_comm_cv (comm_cv IN OUT commcurtype, choice INT);
```

```
    END comm_data;

CREATE OR REPLACE PACKAGE BODY comm_data AS
  PROCEDURE open_comm_cv (comm_cv IN OUT commcurtype, choice INT) IS
  BEGIN
    IF choice = 1 THEN
      OPEN comm_cv FOR SELECT *
      FROM comments
      WHERE feeling = '好评';
    ELSIF choice = 2 THEN
      OPEN comm_cv FOR SELECT *
      FROM comments
      WHERE feeling = '中评';
    ELSIF choice = 3 THEN
      OPEN comm_cv FOR SELECT *
      FROM comments
      WHERE feeling = '差评';
    END IF;
  END;
END comm_data;
```

4.4 批量处理

当在 PL/SQL 中执行含有 SQL 语句的代码块时，PL/SQL 执行引擎只会处理过程语句，把 SQL 语句转交给 SQL 引擎来执行。在 PL/SQL 引擎和 SQL 引擎之间的控制转移叫作上下文切换。每次切换发生时，都会有额外的开销。在大量 SQL 语句的代码块中，由于切换次数过多，会导致性能下降。PL/SQL 的 FORALL 语句是解决这个问题的有效方法，它可以减少 PL/SQL 引擎和 SQL 引擎之间的通信次数，让 PL/SQL 更有效地把多个上下文切换压缩成一个切换，从而提升过程语句的性能。

FORALL 语句会对插入（insert）、更新（update）、合并（merge）以及删除（delete）操作（统称为 DML）进行加速处理。FORALL 命令 PL/SQL 引擎先把一个或多个集合的所有成员绑定到 SQL 语句中，然后再把语句发送给 SQL 引擎，FORALL 是 PL/SQL 语言中最重要的优化特性。

1. FORALL 语句的语法

尽管 FORALL 语句带有一个迭代模式，但它不是一个 FOR 循环，因此既不需要 LOOP，也不需要 END LOOP 语句。语法格式如下。

```
FORALL index IN
[ lower_bound ... upper_bound l
NDICES OF indexing_collection l
VALUES OF indexing_collection
  ]
[SAVE EXCEPTIONS ]
sql_statement;
```

FORALL 语句的参数如表 4.3 所示。

表 4.3　FORALL 语句参数

参　　数	说　　明
index	是一个整数
lower_bound	操作开始的索引值
upper_bound	操作结束的索引值
indexing_collection	指向 sql_statement 所使用绑定数组索引的集合
SAVE EXCEPTIONS	可选的子句,将执行过程中的任何异常保存下来
sql_statement	将对每一个集合元素执行的 SQL 语句

2. FORALL 语句与 FOR LOOP 语句比较

FORALL 语句通常比等效的 FOR LOOP 语句快得多。但是一个 FOR LOOP 语句可以包含多个 DML 语句,而一个 FORALL 语句只能包含一个。FORALL 语句发送给 SQL 引擎的一批 DML 语句的不同之处仅在于它们的 VALUE 和 WHERE 子句。

示例 4.33:对比 FORALL 和 FOR LOOP 的相同执行逻辑性能,可以发现 FORALL 语句带来性能上的提升。

程序代码如下。

```
DECLARE
  TYPE IdTab IS TABLE OF customers_test.custid%TYPE INDEX BY PLS_INTEGER;
  TYPE NameTab IS TABLE OF customers_test.custname%TYPE INDEX BY PLS_INTEGER;
  id    IDTAB;
  name  NAMETAB;
  i CONSTANT NUMBER := 50000;
  t1  TIMESTAMP;
  t2  TIMESTAMP;
  t3  TIMESTAMP;
BEGIN
  FOR j IN 1..i LOOP   --name collections
    id(j) := j;
    name(j) := 'customer' || TO_CHAR(j);
  END LOOP;
    t1 := clock_timestamp();

  FOR i IN 1..i LOOP
    INSERT INTO customers_test VALUES (id(i), name(i));
  END LOOP;

  t2 := clock_timestamp();

  FORALL i IN 1..i
    INSERT INTO customers_test VALUES (id(i), name(i));
```

```
        t3 := clock_timestamp();

    raise notice 'Execution Time';
    raise notice '--------------------------';
    raise notice 'FOR LOOP: %', t2 - t1;
    raise notice 'FORALL  : %', t3 - t2;
    COMMIT;
END;
```

程序运行结果如下。

```
Execution Time
--------------------------
FOR LOOP: 00:00:00.310727
FORALL  : 00:00:00.000205
```

通过结果可以看出 FORALL 语句在性能上的优化。因此，在对性能要求比较高的场景中，应该尽量使用 FORALL 语句来减小批量 DML 语句的执行时间。在接下来的篇幅中会解释 FORALL 的所有特性和一些细节，还会提供许多示例。

3. 对稀疏集合使用 FORALL 语句

如果 FORALL 语句的 bounds 子句引用稀疏集合，则使用 INDICES OF 或 VALUES OF 子句仅指定现有索引值。可以对任何集合使用 INDICES OF，除了以字符串为索引的关联数组。由 PLS_INTEGER 索引的 PLS_INTEGER 元素的集合可以是索引集合，也就是指向另一个集合（索引集合）元素的指针集合。索引集合对于使用不同 FORALL 语句处理同一集合的不同子集很有用。与其将原始集合的元素复制到表示子集的新集合中（这可能会占用大量时间和内存），不如用索引集合表示每个子集，然后在不同 FORALL 语句的 VALUES OF 子句中使用每个索引集合。

示例 4.34：使用带有 INDICES OF 子句的 FORALL 语句修改指定商品的价格。

程序代码如下。

```
declare
  type good_id is table of goods.goodid%type
    index by pls_integer;
  goods_ids   good_id;
  type boolean_aat is table of boolean
    index by pls_integer;
  goods_ids_indices boolean_aat;
begin
  goods_ids(1) := 12560557;
  goods_ids(100) := 9566;
  goods_ids(500) := 41480003669;

  goods_ids_indices(1) := true;
  goods_ids_indices(500) := true;
  goods_ids_indices(700) := true;
```

```
    forall good_index in indices of goods_ids_indices
      between 1 and 500
        update "goods" g set "price" = 200::money
          where g."goodid" = goods_ids(good_index);
end
```

执行结束,查看 goods 表中价格为 200 的商品,程序代码如下。

```
select * from "goods" g where g."price" = 200::money
```

程序运行结果如下。

```
  goodid     |price  |
-----------+-------+
   12560557 |$200.00|
 41480003669|$200.00|
```

从本例中可以看出商品 ID 为 9566 的商品价格并没有发生改变,这是因为从 goods_ids_indices 集合中取得的索引值只有 1 和 500,在 goods_ids 中对应的商品 ID 为 12560557 和 41480003669,因此更新价格的商品只有 2 个。

示例 4.35:使用带有 VALUES OF 子句的 FORALL 语句修改指定商品的价格。

程序代码如下。

```
declare
  type good_id is table of goods.goodid%type
    index by pls_integer;
  goods_ids   good_id;
  type values_aat is table of pls_integer
    index by pls_integer;
  goods_ids_values values_aat;
begin
  goods_ids(1) := 12560557;
  goods_ids(100) := 9566;
  goods_ids(500) := 41480003669;
  
  goods_ids_values(333) := 1;
  goods_ids_values(444) := 100;
  goods_ids_values(555) := 500;
  
  forall good_values in values of goods_ids_values
    update "goods" g set "price" = 200::money
      where g."goodid" = goods_ids(good_values);
end
```

执行结束,查看 goods 表中价格为 200 的商品,程序代码如下。

```
select * from "goods" g where g."price" = 200::money
```

程序运行结果如下。

```
 goodid     |price   |
------------+--------+
   12560557 |$200.00 |
       9566 |$200.00 |
41480003669 |$200.00 |
```

在本例中，商品 ID 的信息存储在 goods_ids 集合中，通过 VALUES OF 子句遍历 good_values 集合中的元素，即对应 goods_ids 的索引，拿到商品的 ID 信息，再对指定商品价格进行更新。

4. FORALL 语句中的异常

FORALL 语句能够以批量的形式给 SQL 引擎传递多个 SQL 语句，这意味着只需要一次上下文切换。不过，对每个语句来说，在 SQL 引擎中还是单独执行的。如果这些 DML 语句中出现了异常，将会导致以下情况：

（1）引发异常的 DML 语句会回滚到这个语句执行之前由 PL/SQL 引擎标识的一个隐式保存点，这个语句已经修改的全部行都会回滚。

（2）这个 FORALL 语句中任何之前已经完成的 DML 操作不会回滚。

（3）如果没有采取特殊操作（给 FORALL 加上 SAVE EXCEPTIONS 子句，稍后讨论），则整个 FORALL 语句停止运行，剩下的语句也不再执行。

以下示例中的 FORALL 语句按此顺序执行这些 DML 语句，除非其中一个引发未处理的异常。

```
DELETE FROM goods WHERE goodid = ids(20);
DELETE FROM goods WHERE goodid = ids(40);
DELETE FROM goods WHERE goodid = ids(60);
```

如果第 3 条语句引发未处理的异常，则 PL/SQL 回滚第 1 条和第 2 条语句所做的更改。如果第 2 条语句引发了未处理的异常，那么 PL/SQL 将回滚第 1 条语句所做的更改，并且从不运行第 3 条语句。

可以通过以下任一方式处理 FORALL 语句中引发的异常。

（1）在引发每个异常时立即处理。

（2）FORALL 语句完成执行后，通过包含 SAVE EXCEPTIONS 子句处理异常。

示例 4.36：在 FORALL 语句执行中发生异常时跳过，接着执行余下的子句。

程序代码如下。

```
DROP TABLE goods;
CREATE TABLE goods (
    goodid NUMBER(4),
    goodname VARCHAR2(12)
);
```

```
CREATE OR REPLACE PROCEDURE p AS
  TYPE NList IS TABLE OF PLS_INTEGER;
  goodids    NList := NList(50, 100, 150);
  errmsg     VARCHAR2(100);
  cook_no    NUMBER;
  good_id    goods.goodid%TYPE;
  good_name  goods.goodname%TYPE;
  err_dml    EXCEPTION;
  PRAGMA EXCEPTION_INIT(err_dml, -10053);
BEGIN
  --Populate table:
  INSERT INTO goods VALUES (50 , 'simba');
  INSERT INTO goods VALUES (100, 'baksha');
  INSERT INTO goods VALUES (150, 'nana');
  COMMIT;

  --Append 9-character string to each goodname:
  FORALL j IN goodids.FIRST..goodids.LAST SAVE EXCEPTIONS
    UPDATE goods SET goodname = goodname || ' RDJCNB'
    WHERE goodid = goodids(j);

  EXCEPTION
    WHEN err_dml THEN
      FOR i IN 1..SQL%BULK_EXCEPTIONS.COUNT LOOP
        errmsg := SQLERRM(-(SQL%BULK_EXCEPTIONS(i).ERROR_CODE));
        raise notice '%', errmsg;

        cook_no := SQL%BULK_EXCEPTIONS(i).ERROR_INDEX;
        raise notice 'cook no:%', cook_no;

        good_id := goodids(cook_no);
        raise notice 'good_id:%', good_id;

        SELECT goodname INTO good_name FROM goods WHERE goodid = good_id;

        raise notice 'goodname:%', good_name;
      END LOOP;
    COMMIT;      --Commit results of successful updates

    WHEN OTHERS THEN
      raise notice 'Unrecognized error.';
      raise notice '%', sqlcode;
      RAISE;
END;

set ora_statement_level_rollback to on;
call p();
select * from "goods" g ;
```

程序运行结果如下。

```
对于可变字符类型来说,值太长了(12)
cook no:2
good_id:100
goodname:baksha
```

查询 goods 表中的数据结果如下。

```
goodid |goodname    |
-------+------------+
   100 |baksha      |
    50 |simba RDJCNB|
   150 |nana RDJCNB |
```

从本例中可以看出,当修改至(100,'baksha')这条数据时,程序打印出相应的异常,并且跳过该行语句,紧接着执行下一条语句。

第 5 章

事 务 处 理

事务是数据库恢复和并发控制的基本单位,具有原子性、一致性、隔离性和持续性特性。KingbaseES 数据库提供了强大的事务模型来支持事务处理,允许多个用户同时处理数据库,并确保每个用户看到一致的数据版本,同时,所有更改都以正确的顺序执行。本章将主要介绍以下内容。
- 事务处理概述。
- 事务处理语句。
- 自治事务。

5.1 事务处理概述

KingbaseES 数据库中的事务是一个最小的执行单元。一个事务可以是一个 SQL 语句,也可以是多个 SQL 语句。一个事务中的语句要么全部执行,要么全不执行。如果所有操作完成,事务将提交,其修改将作用于所有其他数据库进程。如果一个操作失败,事务将回滚,该事务所有操作的影响都将取消。

5.2 事务处理语句

KingbaseES PL/SQL 提供了 COMMIT、ROLLBACK、SET TRANSACTION 以及 LOCK TABLE 等语句。

5.2.1 COMMIT 语句

COMMIT 语句结束当前事务,保存自上次 COMMIT 或 ROLLBACK 以来所有完成的更改,且对其他用户可见,并释放被锁的资源。

5.2.2 ROLLBACK 语句

ROLLBACK 语句结束当前事务,并撤销在该事务期间所作的任何更改,并释放被锁的资源。

如果在此过程中存在错误,例如从表中删除了错误的行,则回滚会恢复原始数据。如果由于 SQL 语句失败或 PL/SQL 引发异常而无法完成事务,则回滚可采取纠正措施,并可能重新开始。

示例 5.1:插入商品类别信息。

注:如果 INSERT 语句试图存储重复的 catgid 主键,PL/SQL 将引发预定义的异常 DUP_VAL_ON_INDEX。为确保撤销对所有操作的更改,异常处理程序运行 ROLLBACK。

程序代码如下。

```
DECLARE new_catgid categories.catgid%TYPE;
BEGIN
  SELECT MAX(catgid) INTO new_catgid FROM categories;

  INSERT INTO categories(catgid,catgname,parentid,currlevel)
  VALUES (new_catgid+1,'手机',new_catgid,1);
  INSERT INTO categories(catgid,catgname,parentid,currlevel)
  VALUES (new_catgid+2,'电脑',new_catgid,2);
  INSERT INTO categories(catgid,catgname,parentid,currlevel)
  VALUES (new_catgid+1,'手机',new_catgid,1);

  COMMIT;
EXCEPTION
  WHEN DUP_VAL_ON_INDEX THEN
    ROLLBACK ;
RAISE NOTICE 'Inserts were rolled back';
END;
```

5.2.3　SET TRANSACTION 语句

SET TRANSACTION 语句为当前事务设置特性,可用的事务特性是事务隔离级别和事务访问模式(只读或者读/写)。SET TRANSACTION 语句必须是事务中的第一条 SQL 语句,并且在事务中只能出现一次。

只读事务对于在其他用户更新同一个表时运行多个查询很有用。在只读事务期间,所有查询都引用数据库的同一个快照,提供多表、多查询、读一致的视图。其他用户可以像往常一样继续查询或更新数据。通过提交或回滚可以结束事务。

如果将事务设置为 READ ONLY,则后续查询只会看到事务开始之前提交的更改。READ ONLY 的使用不会影响其他用户或事务。在只读事务中只允许使用 SELECT、OPEN、FETCH、CLOSE、LOCK TABLE、COMMIT 和 ROLLBACK 语句。SELECT 语句不能是 FOR UPDATE。

示例 5.2:只读事务中的 SET TRANSACTION 语句。

功能描述:只读事务收集商品表中价格在 50 以上、100 以上和 200 以上的商品信息。

程序代码如下。

```
DECLARE
  mt_50 int;
```

```
    mt_100 int;
    mt_200 int;

BEGIN
    COMMIT; -- end previous transaction
    SET TRANSACTION READ ONLY ;

    SELECT count(*)
    INTO mt_50
    FROM goods
    WHERE price>'50.0'::money ;

    SELECT count(*)
    INTO mt_100
    FROM goods
    WHERE price>'100.0'::money ;

    SELECT count(*)
    INTO mt_200
    FROM goods
    WHERE price>'200.0'::money ;

    COMMIT; -- end previous TRANSACTION
END;
```

5.3 自治事务

自治事务是由主事务启动的另一个独立事务。自治事务执行 SQL 操作,并提交或回滚,而不提交或回滚主事务。二者之间的控制流程如图 5.1 所示。

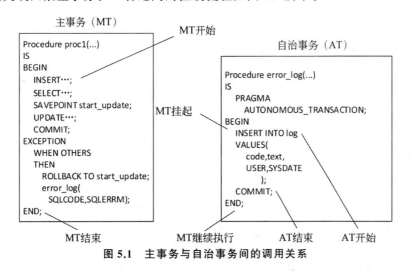

图 5.1 主事务与自治事务间的调用关系

自治事务的优势如下。

（1）启动后，自治事务是完全独立的。它与主事务不共享锁、资源或提交依赖项。即使主事务回滚，也可以记录事件、增加重试计数器等。

（2）自治事务有助于构建模块化、可重用的软件组件。可以将自治事务封装在存储的子程序中。应用程序不需要知道自治事务操作是否成功。

关于自治事务的上下文，主事务与嵌套例程共享其上下文，但不与自治事务共享。当一个自治事务调用另一个（或递归调用自身）时，这些例程不共享事务上下文。当自治事务调用非自治事务时，这些例程共享相同的事务上下文。

关于自治事务是否被调用，可以通过如下两个例子对比。

在 PL/SQL 中，例如匿名块中调用 p1() 的场景，外部块中的 update 会被 p1 中的 COMMIT 提交掉，即使外部块最后有 ROLLBACK，也不会回滚外部块中的事务。

示例 5.3：外部块调用存储过程。

程序代码如下。

```
CREATE OR REPLACE PROCEDURE p1()
AS
BEGIN
  UPDATE customers
  SET email='lin@qq.com'
  WHERE custid=1;
  COMMIT ;
END;

BEGIN
  UPDATE customers
  SET email='jiang@foxmail.com'
  WHERE custid=2;
  p1();
    ROLLBACK;
END;
```

通过如下查询语句可以获取事务执行结果。

```
SELECT custid,email
FROM customers
WHERE custid=1 OR custid=2
```

程序运行结果如下。

```
custid | email |
-------+-------+
   1   | lin@qq.com|
   2   | jiang@foxmail.com|
```

根本原因在于，p1() 内部的事务控制语句影响了外部调用者。而 KingbaseES 提供的自治事务功能通过 PRAGMA AUTONOMOUS_TRANSACTION 语句将 p1() 的事务控制语句完全独立出来,和外部块没有任何关系。

示例 5.4:外部块调用声明为自治事务的存储过程。

程序代码如下。

```
CREATE OR REPLACE PROCEDURE p1()
AS
PRAGMA AUTONOMOUS_TRANSACTION; -- 自治事务定义,表示当前块为自治事务
BEGIN
  UPDATE customers
  SET email='lin2@qq.com'
  WHERE custid=1;
  COMMIT ;
END;

BEGIN
  UPDATE customers
  SET email='jiang2@foxmail.com'
  WHERE custid=2;
  p1();
  ROLLBACK;
END;

  SELECT custid,email
  FROM customers
  WHERE custid=1 OR custid=2
```

程序运行结果如下。

```
custid | email |
------+-----+
    1 |  lin2@qq.com|
    2 |  jiang@foxmail.com |
```

5.3.1 声明自治事务

将一个 PL/SQL 块定义为自治事务,只需要在声明部分包含该语句,语法格式如下。

```
PRAGMA AUTONOMOUS_TRANSACTION;
```

PL/SQL 块可以是如下的任意一种。

(1) 最顶层的(不是嵌套的)匿名块。

(2) 函数或过程,在一个包里或作为独立子程序定义。

在声明中添加自治事务标识是很容易的,但是使用自治事务有一些规则和限制,即不能将 PRAGMA 声明应用于整个包,而是需要在包主体中单独标识每个子程序。

示例 5.5：在包中声明自治函数。

程序代码如下。

```
-- package specification
CREATE OR REPLACE PACKAGE customer_actions AUTHID DEFINER AS
  FUNCTION change_email (cust_id NUMBER, new_email VARCHAR) RETURN VARCHAR;
END customer_actions;

CREATE OR REPLACE PACKAGE BODY customer_actions AS   --package body
  --code for function change_email
  FUNCTION change_email (cust_id NUMBER, new_email VARCHAR)
  RETURN VARCHAR IS
    PRAGMA AUTONOMOUS_TRANSACTION;
    changed_email customers.email%TYPE;
  BEGIN
    UPDATE customers
    SET email = new_email
    WHERE custid= cust_id;

    COMMIT;
    SELECT email INTO changed_email FROM customers
    WHERE custid= cust_id;
    RETURN changed_email;
  END change_email;
END customer_actions;
```

示例 5.6：声明自治子程序。

程序代码如下。

```
CREATE OR REPLACE PROCEDURE change_mobile (cust_id NUMBER, new_moblie VARCHAR)
AS
  PRAGMA AUTONOMOUS_TRANSACTION;
BEGIN
  UPDATE customers
  SET mobile = new_moblie
  WHERE custid= cust_id;

  COMMIT;
END change_mobile;
```

示例 5.7：声明自治 PL/SQL 块。

注：此示例将 PL/SQL 块标记为自治（嵌套的 PL/SQL 块不能是自治的）。

程序代码如下。

```
DECLARE
  PRAGMA AUTONOMOUS_TRANSACTION;
  cust_id INT         := 1;
  new_moblie VARCHAR := '12212212222';
BEGIN
```

```
    UPDATE customers
    SET mobile = new_moblie
    WHERE custid = cust_id;

    COMMIT;
END;
```

5.3.2 从 SQL 中调用自治函数

当需要从事务上下文中隔离一个模块中所作的更改时,应该把该模块定义为自治事务。记录操作日志是自治事务的一个常见的应用。

示例 5.8:调用自治函数。

注:包函数 log_msg 是自治的。因此,当查询调用该函数时,将 1 条消息插入数据库表的 dbg 中,而不会违反写数据库状态的规则(修改数据库表)。

程序代码如下。

```
DROP TABLE IF EXISTS dbg;
CREATE TABLE dbg (msg TEXT);

CREATE OR REPLACE PACKAGE pkg AUTHID DEFINER AS
    FUNCTION log_msg (msg TEXT) RETURN TEXT;
END pkg;

CREATE OR REPLACE PACKAGE BODY pkg AS
    FUNCTION log_msg (msg TEXT) RETURN TEXT IS
        PRAGMA AUTONOMOUS_TRANSACTION;
    BEGIN
        INSERT INTO dbg VALUES (msg);
        RETURN msg;
    END;
END pkg;

-- Invoke package function from query
DECLARE
    my_cust_id      INTEGER;
    my_cust_name    TEXT;
BEGIN
    my_cust_id := 1;

    SELECT pkg.log_msg(custname)
    INTO my_cust_name
    FROM customers
    WHERE custid = my_cust_id;

    ROLLBACK;
END;
```

第 6 章

动态 SQL 语句

动态 SQL 语句是指在执行时进行解析的 SQL 语句,通常 PL/SQL 程序中只执行静态的 SQL 语句。为了支持 PL/SQL 中动态 SQL 语句的执行,KingbaseES 提供了 Native dynamic SQL 和 DBMS_SQL 包两种技术。本章将主要介绍以下内容。

- 动态 SQL 语句概述。
- Native dynamic SQL。
- DBMS_SQL 包。
- SQL 注入。

6.1 动态 SQL 语句概述

1. 动态 SQL 语句的必要性

由于 PL/SQL 程序的执行是在编译阶段对变量进行绑定,识别程序中标识符的位置,检查用户权限、数据库对象等信息,因此在 PL/SQL 中可以直接执行静态 SQL 语句。

但是,编写 SQL 语句时经常需要根据程序运行、客户选择等需求来决定要操作的数据库对象,包括创建表、创建用户、为用户授权,这些操作都是在 SQL 语句运行时完成的。因此,类似 SQL 语句是无法通过编译器编译的,即静态 SQL 语句无法满足这类要求。为了在 PL/SQL 中支持 DDL 语句、DCL 语句以及更加灵活的 SQL 语句,在 PL/SQL 中引入动态 SQL 语句技术。

示例 6.1:创建 1 个存储过程,根据参数指定的列名和值查询店铺信息(使用静态 SQL 语句)。

程序代码如下。

```
CREATE OR REPLACE PROCEDURE dyn_sql_test(
  g_col VARCHAR2,
  g_value VARCHAR2)
AS
  g_name goods.goodname%TYPE;

BEGIN
  SELECT goodname INTO g_name FROM goods WHERE g_col=g_value;
  DBMS_OUTPUT.PUT_LINE(g_name);
END;
```

调用该存储过程时,将发生编译错误,程序代码如下。

```
BEGIN
  dyn_sql_test('good_no','150');
END;
SQL 错误 [42601]: 错误: 语法错误 在 "dyn_sql_test" 或附近的
  Position: 9 At Line: 2, Line Position: 2
```

由该示例可以看出,静态 SQL 语句不具有在运行时提供数据库标识符(如表名、视图名、列名等)的灵活性。为了满足动态查询需求,可以使用动态 SQL 语句。

示例 6.2:创建 1 个存储过程,根据参数指定的列名和值查询店铺信息(使用动态 SQL 语句)。

程序代码如下。

```
CREATE OR REPLACE PROCEDURE dyn_sql_test(
  g_col VARCHAR2,
  g_value VARCHAR2)
AS
  g_name goods.goodname%TYPE;
  g_str varchar2;
BEGIN
  g_str:='SELECT goodname FROM goods WHERE '||g_col||'='||''''||g_value||'''';
  EXECUTE IMMEDIATE g_str INTO g_name;
  RAISE NOTICE '%',g_name;
END;
```

此时,可以根据不同的输入参数进行动态查询,程序代码如下。

```
BEGIN
  dyn_sql_test('goodid','7532692');
  dyn_sql_test('model','SIS97915342');
END;
```

程序运行结果如下。

```
NOTICE:小迷糊素颜霜 20g
NOTICE:知识图谱:方法、实践与应用
```

动态 SQL 是一种用于在运行时生成和运行 SQL 语句的编程方法,将包含不确定数据库对象、创建数据库对象等的 SQL 语句封装成一个字符串,在编译阶段只进行字符串的语法检查,在运行时才进行 SQL 语句的分析与执行。通常,动态 SQL 被广泛应用于如下场景中。

(1) 编写诸如临时查询类的通用且灵活的程序。
(2) 编写必须运行数据库定义语言(DDL)语句的程序。
(3) 在编译时不知道 SQL 语句的全文、编号或其输入和输出变量的数据类型。

2. 动态 SQL 的编写方法

PL/SQL 提供了以下两种编写动态 SQL 的方法。

（1）Native dynamic SQL，一种 PL/SQL 语言（即本机）的功能，用于构建和运行动态 SQL 语句。

（2）DBMS_SQL 包，用于构建、运行和描述动态 SQL 语句的 API。

Native dynamic SQL 代码比使用 DBMS_SQL 包的等效代码更易于读写，且运行速度更快（尤其是当它可以被编译器优化时）。

但是，要编写 Native dynamic SQL 代码，必须在编译时知道动态 SQL 语句输入和输出变量的数量和数据类型。如果在编译时不知道此信息，则必须使用 DBMS_SQL 包。

如果希望存储的子程序隐式返回查询结果（而不是通过 OUT REFCURSOR 参数），也必须使用 DBMS_SQL 包。

当需要 DBMS_SQL 包和 Native dynamic SQL 时，可以使用 DBMS_SQL.TO_REFCURSOR 函数和 DBMS_SQL.TO_CURSOR_NUMBER 函数，在它们之间切换。

3. 动态 SQL 与静态 SQL 语句的比较

静态 SQL 语句虽然在 PL/SQL 程序中有广泛使用，但在应用程序开发过程的某些情况下，只能使用动态 SQL 语句而不能使用静态 SQL 语句。

（1）在程序编译阶段，不能提供完整的 SQL 语句。在一些复杂应用中，需要根据用户的选择或输入决定数据的查询，即用户的查询条件在编译阶段是不确定的，只有在执行时才能知道，此时只能使用动态 SQL 语句，而不能使用静态 SQL 语句。

（2）PL/SQL 中的静态 SQL 语句包括查询语句（SELECT）、DML 语句（INSERT、UPDATE、DELETE、MERGE），事务控制语句（COMMIT、ROLLBACK、SAVEPOINT、SET TRANSACTION）以及表锁定语句（LOCK TABLE）。如果要在 PL/SQL 中执行其他语句，如 DDL 语句、DCL 语句、会话控制语句、系统控制语句等，只能使用动态 SQL 语句。

虽然动态 SQL 语句可以让开发人员在运行时动态切换表名、列名等数据库标识符，以及在 PL/SQL 中执行各种 DDL 操作、DCL 操作，但是在下列几种情况下，静态 SQL 语句更具优势。

（1）静态 SQL 语句在编译时验证 SQL 语句引用的数据库对象是否有效，如果依赖的对象不存在或失效，那么 SQL 语句编译错误。动态 SQL 语句只有在运行时才能知道类似的错误。

（2）静态 SQL 语句在编译时检查用户是否具有相应数据库对象的访问权限，如果用户没有权限，则编译失败。动态 SQL 语句对用户的权限检查在运行时进行。

（3）使用静态 SQL 语句可以对要执行的 SQL 语句进行性能优化调整，提高程序的执行性能。但动态 SQL 语句无法进行优化处理。

6.2 Native dynamic SQL

Native dynamic SQL 与 DBMS_SQL 包相比更灵活，执行效率更高，Native dynamic SQL 为动态非查询语句、动态单行查询语句、动态多行查询语句提供了下列 3 种运行模式。

(1) 使用 EXECUTE IMMEDIATE 语句处理大多数语句,包括非查询语句和单行查询语句。

(2) 使用 OPEN FOR、FETCH 和 CLOSE 语句实现多行查询。

(3) 使用带 BULK COLLECT INTO 子句的 EXECUTE IMMEDIATE 语句实现多行查询,即动态批处理绑定。

6.2.1 EXECUTE IMMEDIATE 语句

EXECUTE IMMEDIATE 语句是 Native dynamic SQL 处理大多数动态 SQL 语句的方法。该语句的语法格式如下。

```
EXECUTE IMMEDIATE dynamic_SQL_string
    [ [ BULK COLLECT] INTO {define_variable[, define_variable] ... | record}]
    [USING [IN | OUT | IN OUT] bind_argument
        [, [ IN | OUT | IN OUT] bind_argumen t] ... ];
```

其中:

(1) dynamic_SQL_string:是包含 SQL 语句或 PL/SQL 块的字符串。该字符串中,数据库对象名使用变量表示,值使用占位符表示。

(2) INTO 字句:指定接收单行查询结果的 OUT 模式的绑定变量。如果 BULK COLLECT 作为 INTO 的开始,则是动态批处理绑定,将多行返回结果保存到集合中。其中 define_variable 是接收由查询返回的列值的变量;record 是用户自定义类型或基于% ROWTYPE 的记录,接收由查询返回的整行的值。

(3) USING 字句:为 SQL 字符串提供绑定参数,可用于动态 SQL 和动态 PL/SQL。如果动态 SQL 语句中带有 RETURNING 子句,则绑定变量可以是 IN 模式,向 SQL 语句传值;可以是 OUT 模式,接收返回的值;也可以是 IN OUT 模式。如果动态 SQL 语句不带 RETURNING 子句,则绑定变量只能是 IN 模式。

如果动态 SQL 语句不含和绑定变量相关联的占位符,并且不返回结果集,这样的 EXECUTE IMMEDIATE 语句则不需要其他子句。

示例 6.3:动态 DDL 操作。

功能描述:查询 TEST 表是否存在,如果不存在,则创建该表。

程序代码如下。

```
DECLARE
  i NUMBER;
  v_str VARCHAR2(100);
BEGIN
  SELECT count(*)INTO i FROM user_tables
  WHERE table_name ='TEST';
IF i=0 THEN
  v_str:='CREATE TABLE TEST(id NUMBER,name NVARCHAR)';
  EXECUTE IMMEDIATE v_str;
  END IF;
END;
```

除了在匿名块中使用,命名块中也可以使用动态 DDL 语句。

示例 6.4:存储过程中执行动态 DDL 操作。

功能描述:创建一个存储过程,根据参数指定列名和数据类型动态创建表。

程序代码如下。

```
CREATE OR REPLACE PROCEDURE create_table(
  p_tablename VARCHAR2,
  p_col1 VARCHAR2, p_col1_type VARCHAR2,
  p_col2 VARCHAR2, p_col2_type VARCHAR2)
AS
  v_creation VARCHAR2(100);
BEGIN
  cocreation:='CREATE TABLE '||p_tablename||'('|| p_col1||' '||p_col1_type||' primary key,'||p_col2||' '||p_col2_type||')';
EXECUTE IMMEDIATE v_creation;
END;
```

调用该存储过程的代码如下。

```
BEGIN
    create_table('TEST1','id','NUMBER','name','NVARCHAR');
END;
```

示例 6.5:动态 DCL 操作。

功能描述:创建一个存储过程,根据参数指定用户名称和权限名称,动态给用户授权。

程序代码如下。

```
CREATE OR REPLACE PROCEDURE grant_priv(
    p_priv VARCHAR2,
    p_scheme VARCHAR2,
    p_username VARCHAR2)
AS
    sql_str varchar2(100);
BEGIN
    sql_str:='GRANT '||p_priv ||' ON SCHEMA '||p_scheme||' TO '||p_username;
    EXECUTE IMMEDIATE sql_str;
END;
```

调用该存储过程,给特定用户授予特定系统权限。

```
BEGIN
    grant_priv('CREATE','sales','public');
END;
```

在动态 DML 语句中,数据库标识符,如表名、列名等通常使用 PL/SQL 变量或者参数来标识,而与绑定变量相关的动态值使用占位符(冒号加标识符)来表示。

示例 6.6：不带占位符的动态 DML 语句。

功能描述：修改 custid 为 1 的客户邮箱。

程序代码如下。

```
CREATE OR REPLACE PROCEDURE update_email(
  table_name VARCHAR2)
AS
    sql_str varchar2(100);
BEGIN
    sql_str:='UPDATE '||table_name||'
      SET EMAIL=''jao@qq.com''
        WHERE custid=1 ';
    EXECUTE IMMEDIATE sql_str;
END;
BEGIN
  update_email('customers');
END;
```

示例 6.6 的 DML 语句中唯一的变量是表名，没有使用占位符来表示动态值，则不需要使用 USING 语句来指定绑定变量。

把之前 DML 语句中硬编码的值用占位符来表示，如示例 6.7 所示。

示例 6.7：带有占位符的动态 DML 语句。

功能描述：修改指定用户的邮箱。

程序代码如下。

```
CREATE OR REPLACE PROCEDURE update_email(
    table_name VARCHAR2,
    new_email VARCHAR2,
    custid VARCHAR2)
IS
    v_str VARCHAR2(100);
BEGIN
    v_str:='UPDATE '||table_name||'
      SET EMAIL=:EMAIL
        WHERE custid=:custid ';
EXECUTE IMMEDIATE v_str USING new_email,custid;
END;
BEGIN
  update_email('customers','joa@qq.com','1');
END;
```

在 DML 语句的操作过程中获取数据也非常重要，RETURNING 子句可以从修改的行中返回数据，避免执行额外的数据库查询来收集数据，并且在难以识别修改的行时尤其有用。

对于包含 RETURNING 子句的动态 DML 语句，需要通过 RETURNING INTO 子句或 USING 子句指定接收返回结果的绑定变量。如果是使用 USING 子句指定返回结果的绑定变量,那么绑定变量必须是 OUT 模式或 IN OUT 模式。

示例 6.8:包含 RETURNING 子句的动态 DML 语句。

功能描述:创建一个存储过程,删除指定记录,同时使用 RETURNING 子句返回操作结果。

程序代码如下。

```
CREATE OR REPLACE PROCEDURE delete_customer(
    p_col VARCHAR2,
    p_val VARCHAR2)
AS
    v_str VARCHAR2(200);
    v_name customers.custname%TYPE;
    v_gender customers.gender%TYPE;
BEGIN
    v_str:='DELETE FROM customers
        WHERE '||p_col||' =:p_val
        RETURNING custname,gender INTO :v_name,:v_gender';
    EXECUTE IMMEDIATE v_str USING p_val RETURNING INTO v_name,v_gender;
    RAISE NOTICE '% %',v_name,v_gender;
END;
```

之后调用该存储过程,删除记录并返回删除记录的部分信息。

```
BEGIN
    delete_customer('custid','200');
END;
```

程序运行结果如下。

```
NOTICE:于光明 女
```

示例 6.9:包含 RETURNING 子句的动态 DML 语句,使用 USING 子句接收返回结果。

功能描述:创建 1 个存储过程,删除指定记录,使用 USING 子句指定 OUT 模式的绑定变量,用来接收 DML 语句的返回结果。

程序代码如下。

```
CREATE OR REPLACE PROCEDURE delete_customer2(
    p_col VARCHAR2,
    p_val VARCHAR2)
AS
    v_str VARCHAR2(200);
    v_name customers.custname%TYPE;
    v_gender customers.gender%TYPE;
BEGIN
    v_str:='DELETE FROM customers
        WHERE '||p_col||' =:p_val
        RETURNING custname,gender INTO :v_name,:v_gender';
```

```
        EXECUTE IMMEDIATE v_str USING p_val,out v_name,out v_gender;
        RAISE NOTICE '% %',v_name,v_gender;
END;
```

调用该存储过程,删除记录,并返回删除记录的部分信息。

```
BEGIN
    delete_customer2('custid','201');
END;
```

程序运行结果如下。

```
NOTICE:易博裕 女
```

使用 EXECUTE IMMEDIATE 执行单行查询的动态 SQL 语句,可能是最常用到的动态 SQL,它是一种有固定数量的绑定变量的查询,需要使用 INTO 语句指定绑定变量,接收查询结果,绑定变量的模式默认为 OUT。

示例 6.10：动态单行查询操作。

功能描述：创建 1 个存储过程,删除指定记录,使用 USING 子句指定 OUT 模式的绑定变量,用来接收 DML 语句的返回结果。

程序代码如下。

```
CREATE OR REPLACE PROCEDURE query_customer(
    p_col VARCHAR2,
    p_val VARCHAR2)
IS
    v_name customers.custname%TYPE;
    v_gender customers.gender%TYPE;
    v_str VARCHAR2(200);
BEGIN
    v_str:='SELECT custname,gender
        FROM customers
        WHERE '||p_col||' =:p_val';
    EXECUTE IMMEDIATE v_str INTO v_name,v_gender USING p_val;
    RAISE NOTICE '% %',v_name,v_gender;
END;
```

调用该存储过程,代码如下。

```
BEGIN
   query_customer('custid','202');
   query_customer('custname','邹光临');
END;
```

程序运行结果如下。

```
NOTICE:郭咏歌 女
NOTICE:邹光临 女
```

动态 SQL 中还可以调用子程序,需要注意的是,在动态 SQL 语句中,子程序参数使用占位符来表示;执行动态 SQL 语句时,使用 USING 语句指定与占位符对应的绑定变量。

示例 6.11:在动态 SQL 中调用子程序。

功能描述:创建一个存储过程,在动态 SQL 语句中调用它。

程序代码如下。

```
CREATE OR REPLACE PROCEDURE create_category(
  catgid IN OUT NUMBER,
  catgname IN VARCHAR2,
  parentid IN NUMBER,
  currlevel IN NUMBER)
AS
BEGIN
  INSERT INTO categories(catgid, catgname, parentid, currlevel)
  VALUES (catgid,catgname,parentid,currlevel);
END;
```

使用动态 SQL 调用该存储过程。

```
    DECLARE
    plsql_block VARCHAR2;
    catgid NUMBER;
    catgname VARCHAR2;
    parentid NUMBER;
    currlevel NUMBER;
BEGIN
    catgid:=3000;
    catgname:='吹风机';
    parentid:=186;
    currlevel:=1;
    plsql_block := 'BEGIN create_category(:a,:b,:c,:d); END;';
     EXECUTE  IMMEDIATE  plsql _ block  using  INOUT  catgid, catgname, parentid,
currlevel;
END;
```

6.2.2　OPEN FOR、FETCH 和 CLOSE 语句

OPEN FOR、FETCH 和 CLOSE 是动态 SQL 用来处理多行查询,把游标变量和动态 SQL 语句相关联的语句,操作步骤与静态游标类似,基本步骤如下。

(1) 定义游标引用类型。

(2) 定义游标变量。

(3) 构建动态查询字符串。

(4) 使用 OPEN FOR 语句将游标变量与动态 SQL 语句相关联。在 OPEN FOR 语句的 USING 子句中,为动态 SQL 语句中的每个占位符指定一个绑定变量。

(5) 使用循环结构 FETCH 语句一次检索一行、多行或所有结果集。

(6) 使用 CLOSE 语句关闭游标变量。

示例 6.12：带有 OPEN FOR、FETCH 和 CLOSE 语句的 Native dynamic SQL。
功能描述：本例中列出前 5 条女性顾客信息，一次检索一行结果集。
程序代码如下。

```
DECLARE
  TYPE CustCurTyp IS REF CURSOR;
  v_cust_cursor   CustCurTyp;
  cust_record   customers%ROWTYPE;
  v_stmt_str   VARCHAR2(200);
BEGIN
  --动态 SQL 语句中使用占位符
  v_stmt_str := '
  SELECT custid,custname,gender
  FROM customers
  WHERE gender = :j
  LIMIT 5';

  --打开游标为占位符绑定 in 参数
  OPEN v_cust_cursor FOR v_stmt_str USING '女';

  --每次从结果集中取出一行数据
  LOOP
    FETCH v_cust_cursor INTO cust_record;
      EXIT WHEN v_cust_cursor%NOTFOUND;
        RAISE NOTICE 'cust_record: %', cust_record;
  END LOOP;

  --关闭游标
  CLOSE v_cust_cursor;
END;
```

程序运行结果如下。

```
NOTICE:cust_record: (2,江文曜,"女 ",,,,,,,,,)
NOTICE:cust_record: (3,吕浩初,"女 ",,,,,,,,,)
NOTICE:cust_record: (4,萧承望,"女 ",,,,,,,,)
NOTICE:cust_record: (5,程同光,"女 ",,,,,,,,)
NOTICE:cust_record: (6,邱涛,"女 ",,,,,,,,)
```

6.2.3 重复的占位符名称

在前面几个例子中，有的在 Native dynamic SQL 中使用绑定变量的用法。动态 SQL 语句中可以使用绑定变量的，可以将动态 SQL 语句中的占位符替换为数值的直接量（文本、变量、表达式）；对于不可以使用占位符的有模式元素的名称（表名、列名等），则必须使用拼接语句。

在动态构造和执行的 SQL 语句中，Native dynamic SQL 通过位置将占位符和 USING 子句的绑定参数关联，而非通过名称。所以，可能存在多个拥有相同名称的占位符，在这种情况下,占位符和绑定变量的关联关系取决于动态 SQL 语句的类型,参照以下规则。

1. 动态 SQL 语句不是匿名块或 CALL 语句

如果动态 SQL 语句不是匿名 PL/SQL 块或 CALL 语句,则占位符名称重复无关紧要。占位符在 USING 子句中按位置而不是按名称与绑定变量相关联。

例如,在如下动态 SQL 语句中,名称:x 的重复是无关紧要的。

```
sql_stmt := 'INSERT INTO payroll VALUES (:x, :x, :y, :x)';
```

在相应的 USING 子句中,必须提供 4 个绑定变量。它们可以不同,例如:

```
EXECUTE IMMEDIATE sql_stmt USING a, b, c, d;
```

应用前面的 EXECUTE IMMEDIATE 语句可以运行以下 SQL 语句。

```
INSERT INTO payroll VALUES (a, b, c, d)
```

要将相同的绑定变量与每次出现的 :x 关联,必须重复该绑定变量,例如:

```
EXECUTE IMMEDIATE sql_stmt USING a, a, b, a;
```

应用前面的 EXECUTE IMMEDIATE 语句运行以下 SQL 语句,代码如下。

```
INSERT INTO payroll VALUES (a, a, b, a)
```

2. 动态 SQL 语句是匿名块或 CALL 语句

如果动态 SQL 语句表示匿名 PL/SQL 块或 CALL 语句,则占位符名称的重复是有效的。每个唯一的占位符名称在 USING 子句中必须有一个对应的绑定变量。

如果重复占位符名称,则无须重复其对应的绑定变量,对该占位符名称的所有引用都对应于 USING 子句中的一个绑定变量。

示例 6.13:动态 PL/SQL 块中的重复占位符名称。

功能描述:在本例中,第一个唯一占位符名称:x 的所有引用都与 USING 子句中的第一个绑定变量 *a* 相关联,第二个唯一占位符名称:y 与 USING 子句中的第二个绑定变量 *b* 相关联。

程序代码如下。

```
CREATE PROCEDURE add_int (
  a INTEGER,
  b INTEGER,
  c INTEGER,
  d INTEGER )
IS
BEGIN
  RAISE NOTICE 'a + b + c + d = %', a+b+c+d;
END;
```

```
DECLARE
  a INTEGER := 1;
  b INTEGER := 2;
  plsql_block VARCHAR2(100);
BEGIN
  plsql_block := 'BEGIN add_int(:x, :x, :y, :x); END;';
  EXECUTE IMMEDIATE plsql_block USING a, b;
END;
```

程序运行结果如下。

```
NOTICE:  a + b + c + d = 5
```

6.3　DBMS_SQL 包

DBMS_SQL 包定义了一个管理 PL/SQL 游标句柄的实体。因为 PL/SQL 游标句柄是一个会话级全局唯一的标识（整数），所以可以在一个会话中跨调用序列传递并使用它。

如果满足以下任一条件，则必须使用 DBMS_SQL 包运行动态 SQL 语句。

（1）直到运行时才知道 SELECT 列表。

（2）直到运行时才知道必须绑定 SELECT 或 DML 语句中的哪些占位符。

（3）希望存储子程序隐式返回查询结果（而不是通过 OUT REFCURSOR 参数），需要 DBMS_SQL.RETURN_RESULT 过程。

如果是以下情况，必须使用 Native dynamic SQL 而不是 DBMS_SQL 包：

（1）动态 SQL 语句将行检索到记录中。

（2）希望在发出作为 INSERT、UPDATE、DELETE、MERGE 或单行 SELECT 语句的动态 SQL 语句后使用 SQL 游标属性％FOUND、％ISOPEN、％NOTFOUND 或％ROWCOUNT。

6.3.1　DBMS_SQL 包中的常用方法

DBMS_SQL 是 KingbaseES 的 plsql 扩展插件提供的系统包。DBMS_SQL 包提供了一系列的过程和函数，专门用于动态 SQL 的操作，以下简要介绍使用频率较高的几种。

（1）function open_cursor：打开一个新游标，返回游标的 ID。

（2）procedure close_cursor(c in out integer)：关闭一个动态游标，参数为 open_cursor 所打开的游标。

（3）procedure parse(c in integer, statement in varchar2, language_flag in integer)：对动态游标所提供的 SQL 语句进行解析，参数 c 表示游标，statement 为 SQL 语句。

（4）procedure define_column(c in integer, position in integer, column any datatype, [column_size in integer])：定义动态游标所能得到的对应值，其中 c 为动态游标，positon 为对应动态 SQL 中的位置（从 1 开始），column 为该值所对应的变量，可以为任何类型，column_size 只有在 column 为定义长度的类型中使用如 VARCHAR2、CHAR 等参数（该

过程有多种情况,此处只对一般使用到的类型进行表述)。

(5) function execute(c in integer):执行给定的游标,返回已处理的行数(仅对INSERT、UPDATE和DELETE语句,其他类型的语句返回值是不确定的)。

(6) function fetch_rows(c in integer):从给定游标中获取数据,并且返回实际获取的行数。只要还有行需要提取,就可以重复调用FETCH_ROWS。这些行被检索到缓冲区中,如果需要读取数据,则需要调用COLUMN_VALUE函数来读取。不能采用NO_DATA_FOUND或游标属性%NOTFOUND判断是否检索到数据。

(7) procedure column_value(c in integer, position in integer, value):用于访问给定游标、给定位置,指定为基本数据类型的列值。此过程用于获取fetch_rows调用后的数据。

(8) procedure bind_variable(c in integer, name in varchar2, value):定义动态SQL语句(DML)中所对应字段的值,c为游标,name为字段名称,value为字段的值。

6.3.2 DBMS_SQL 包操作流程

对于一般的SELECT操作,如果使用动态SQL语句,步骤如图6.1(a)所示。

对于DML操作(INSERT,UPDATE),步骤如图6.1(b)所示。

对于DELETE操作,步骤如图6.1(c)所示。

对于DDL操作,步骤如图6.1(d)所示。

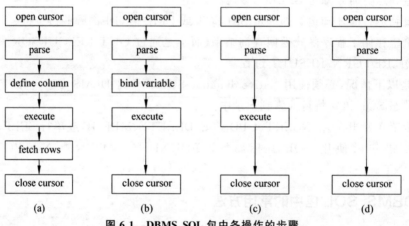

图 6.1 DBMS_SQL 包中各操作的步骤

示例 6.14:动态 DDL 语句。

功能描述:本例演示使用 DBMS_SQL 包执行动态 DDL 语句的过程。

程序代码如下。

```
CREATE OR REPLACE PROCEDURE pro_test_dbms_sql(l_num in number,l_tabname in
varchar2)
IS
  l_cur integer;
  l_sql1 varchar2(400);
  l_sql2 varchar2(400);
BEGIN
  l_cur :=dbms_sql.open_cursor;
```

```
  IF l_num=1
  THEN
    l_sql2 :='drop table '||l_tabname;
    dbms_sql.parse(l_cur,l_sql2,dbms_sql.native);
    dbms_sql.close_cursor(l_cur);
  END IF;

  IF l_num=0
  THEN
    l_sql1 :='create table '||l_tabname||' (id number(10))';
    dbms_sql.parse(l_cur,l_sql1,dbms_sql.native);
    dbms_sql.close_cursor(l_cur);
  END IF;
EXCEPTION
  WHEN OTHERS THEN
RAISE;
END;

EXEC pro_test_dbms_sql(0,'test');
select table_name from user_tables where table_name='TEST';
```

已成功创建 test 表,执行结果如下。

```
table_name
-------+
   TEST
```

之后再执行删除 test 表的操作。

```
exec  pro_test_dbms_sql(1,'test');
select count(1) from user_tables where table_name='TEST';
```

可以看出 test 表已被删除。

```
count
-------+
    0
```

示例 6.15:动态 DML 语句。

功能描述:本例演示使用 DBMS_SQL 包执行动态 DML 语句的过程。计算订单的最终金额,其中,最终金额=商品总金额+运费-折扣金额。

程序代码如下。

```
CREATE OR REPLACE PROCEDURE pro_update_finlbal
AS
  l_num integer;
  l_cur integer;
  l_sql varchar2(400);
  l_return integer default 0;
```

```
BEGIN
SELECT max(custid) into l_num from orders;
for i in 1..l_num   loop
  l_cur :=dbms_sql.open_cursor;
  l_sql:='update orders set finlbal=totlbal+carriage-discamt where custid='||i;

  dbms_sql.parse(l_cur,l_sql,dbms_sql.native);
  l_return :=dbms_sql.execute(l_cur);
COMMIT;
  dbms_sql.close_cursor(l_cur);
END LOOP;
EXCEPTION
WHEN OTHERS THEN
  dbms_sql.close_cursor(l_cur);
  dbms_output.put_line('执行失败');
END;
```

先清除 Orders 表中的最终金额,再调用上述动态 SQL 语句。

```
UPDATE orders SET finlbal=NULL;
EXEC pro_update_finlbal();
SELECT totlbal,carriage,discamt,finlbal FROM orders LIMIT 5;
```

程序运行结果如下。

```
totlbal | carriage | discamt | finlbal |
--------+----------+---------+---------+
$79.10  |$0.00     |$0.02    |$79.08   |
$153.00 |$0.00     |$54.00   |$99.00   |
$98.00  |$0.00     |$0.01    |$97.99   |
$29.90  |$0.00     |$3.00    |$26.90   |
$142.80 |$0.00     |$0.00    |$142.80  |
```

示例 6.16:动态查询语句。

功能描述:本例演示使用 DBMS_SQL 包执行动态查询语句的过程。

程序代码如下。

```
CREATE OR REPLACE PROCEDURE dyn_single_query_sp(
  p_col IN VARCHAR2,
  p_value IN VARCHAR2)
IS
  lv_query LONG;
  lv_status INTEGER;
  lv_cursor INTEGER;
  lv_custid customers.custid%TYPE;
  lv_custname customers.custname%TYPE;
  lv_email customers.email%TYPE;
BEGIN
  lv_cursor:=DBMS_SQL.OPEN_CURSOR;
```

```
  lv_query :='
    SELECT custid,custname,email
    FROM customers WHERE '|| p_col ||' = ' || ':ph_value';
  DBMS_SQL.PARSE(lv_cursor, lv_query, DBMS_SQL.NATIVE);
  DBMS_SQL.BIND_VARIABLE(lv_cursor, ':ph_value',p_value);
  DBMS_SQL.DEFINE_COLUMN(lv_cursor, 1, lv_custid);
  DBMS_SQL.DEFINE_COLUMN(lv_cursor, 2, lv_custname);
  DBMS_SQL.DEFINE_COLUMN(lv_cursor, 3, lv_email);

  lv_status := DBMS_SQL.EXECUTE(lv_cursor);
  IF(DBMS_SQL.FETCH_ROWS(lv_cursor) > 0) THEN
    DBMS_SQL.COLUMN_VALUE(lv_cursor, 1, lv_custid);
    DBMS_SQL.COLUMN_VALUE(lv_cursor, 2, lv_custname);
    DBMS_SQL.COLUMN_VALUE(lv_cursor, 3, lv_email);
    RAISE NOTICE '% % %',lv_custid,lv_custname,lv_email;
  END IF;
  DBMS_SQL.CLOSE_CURSOR(lv_cursor);
END;
```

调用该存储过程,传入指定的列名和参数进行查询。

```
BEGIN
  dyn_single_query_sp('custid','1');
  dyn_single_query_sp('custname','林心水');
END;
```

程序运行结果如下。

```
NOTICE:1 林心水 jao@qq.com
NOTICE:1 林心水 jao@qq.com
```

6.3.3 其他常用 DBMS_SQL 方法

1. DBMS_SQL.RETURN_RESULT 过程

DBMS_SQL.RETURN_RESULT 过程允许存储的子程序将查询结果隐式返回给客户端程序(间接调用子程序)或子程序的直接调用者。DBMS_SQL.RETURN_RESULT 返回结果后,只有接收方可以访问。

DBMS_SQL.RETURN_RESULT 有以下两个重载。

```
PROCEDURE RETURN_RESULT (rc IN REFCURSOR,
         to_client IN BOOLEAN DEFAULT TRUE);


PROCEDURE RETURN_RESULT (rc IN INTEGER,
         to_client IN BOOLEAN DEFAULT TRUE);
```

其中,rc 参数是打开的游标变量(REFCURSOR),或打开的游标的游标编号

(INTEGER)。要打开游标并获取其游标句柄,请调用 DBMS_SQL.OPEN_CURSOR 函数。

当 to_client 参数为 TRUE(默认)时,DBMS_SQL.RETURN_RESULT 过程将查询结果返回给客户端程序(间接调用子程序);为 FALSE 时,过程将查询结果返回给子程序的直接调用者。

示例 6.17:DBMS_SQL.RETURN_RESULT 过程。

功能描述:在本例中,过程 p 调用 DBMS_SQL.RETURN_RESULT 时没有可选的 to_client 参数(默认为 TRUE)。因此,DBMS_SQL.RETURN_RESULT 将查询结果返回给子程序客户端(调用 p 的匿名块)。在 p 向匿名块返回结果后,只有匿名块可以访问该结果。

程序代码如下。

```
CREATE OR REPLACE PROCEDURE p AUTHID DEFINER AS
  c1 SYS_REFCURSOR;
  c2 SYS_REFCURSOR;
BEGIN
  OPEN c1 FOR
    SELECT custid, custname, gender
    FROM customers
    WHERE gender='男'
  LIMIT 3;

  DBMS_SQL.RETURN_RESULT (c1);
  --Now p cannot access the result.

  OPEN c2 FOR
    SELECT custid, custname, dob
    FROM customers
    WHERE YEAR(dob)=1990
  LIMIT 5;

  DBMS_SQL.RETURN_RESULT (c2);
END;

BEGIN
  p;
END;
```

程序运行结果如下。

```
custid | custname | gender |
-------+----------+--------+---
     1 | 林心水   | 男     |
     7 | 石盈     | 男     |
     8 | 梁豹     | 男     |
```

```
custid | custname | dob|
-------+--------- +---------+
     4 |   萧承望   |1990-08-27 00:00:00   |
    23 |   冯彻     |1990-06-03 00:00:00   |
    27 |   乔御     |1990-09-03 00:00:00   |
    39 |   石豹     |1990-01-07 00:00:00   |
    62 |   郝骄     |1990-05-28 0:00:00    |
```

2. DBMS_SQL.GET_NEXT_RESULT 过程

DBMS_SQL.GET_NEXT_RESULT 过程获取 DBMS_SQL.RETURN_RESULT 过程的下一个结果返回给接收者。这两个过程以相同的顺序返回结果。

DBMS_SQL.GET_NEXT_RESULT 有以下两个重载。

```
PROCEDURE GET_NEXT_RESULT (c IN INTEGER, rc OUT REFCURSOR);
PROCEDURE GET_NEXT_RESULT (c IN INTEGER, rc OUT INTEGER);
```

其中，c 参数是打开游标的游标句柄，该游标直接或间接调用使用 DBMS_SQL.RETURN_RESULT 过程隐式返回查询结果的子程序。

要打开游标并获取其游标句柄，应该调用 DBMS_SQL.OPEN_CURSOR 函数。DBMS_SQL.OPEN_CURSOR 有一个可选参数，即 treat_as_client_for_results。当此参数为 FALSE（默认值）时，打开此游标（以调用子程序）的调用者不被视为从使用 DBMS_SQL.RETURN_RESULT 的子程序接收客户端查询结果的客户端——这些查询结果返回到上层的客户端。当此参数为 TRUE 时，调用者被视为客户端。

示例 6.18：DBMS_SQL.GET_NEXT_RESULT 过程。

功能描述：在本例中，过程 get_student_info 使用 DBMS_SQL.RETURN_RESULT 将两个查询结果返回给客户端程序，并由匿名块动态调用。因为匿名块需要接收 get_student_info 返回的两个查询结果，所以匿名块打开一个游标，以使用 DBMS_SQL.OPEN_CURSOR 调用 get_student_info，并将参数 treat_as_client_for_results 设置为 TRUE。因此，DBMS_SQL.GET_NEXT_RESULT 将其结果返回给匿名块，后者使用游标 rc 来获取它们。

程序代码如下。

```
CREATE OR REPLACE PROCEDURE get_customer_info (custid IN INT) AUTHID DEFINER AS
  rc  SYS_REFCURSOR;
BEGIN
  --返回顾客表的信息
  OPEN rc FOR SELECT custid, custname, gender
      FROM customers
      WHERE custid = custid;
  DBMS_SQL.RETURN_RESULT(rc);
END;

DECLARE
  c     INTEGER;
```

```
    rc  SYS_REFCURSOR;
    n   NUMBER;

    custid    customers.custid%TYPE;
    custname  customers.custname%TYPE;
    gender    customers.gender%TYPE;
BEGIN
    c := DBMS_SQL.OPEN_CURSOR(true);
    DBMS_SQL.PARSE(c, 'BEGIN get_customer_info(:custid); END;', DBMS_SQL.NATIVE);
    DBMS_SQL.BIND_VARIABLE(c, ':custid', 1);
    n := DBMS_SQL.EXECUTE(c);

    --Get customer info
    DBMS_SQL.get_next_result(c, rc);
    FETCH rc INTO custid, custname, gender;
      RAISE NOTICE 'customer: %.% gender: %', custid, custname, gender;

    DBMS_SQL.CLOSE_CURSOR(c);
END main;
```

程序运行结果如下。

```
NOTICE:customer: 1.林心水 gender: 男
```

3. DBMS_SQL.TO_REFCURSOR 函数

DBMS_SQL.TO_REFCURSOR 函数将 SQL 游标句柄转换为弱类型的游标变量，可以在 Native dynamic SQL 语句中使用该变量。

将 SQL 游标句柄传递给 DBMS_SQL.TO_REFCURSOR 函数之前，必须先使用 OPEN、PARSE 和 EXECUTE（否则会发生错误）。

将 SQL 游标句柄转换为 REFCURSOR 变量后，DBMS_SQL 操作只能将其作为 REFCURSOR 变量访问，而不能作为 SQL 游标句柄访问。例如，使用 DBMS_SQL.IS_OPEN 函数来查看转换后的 SQL 游标句柄是否仍然打开，会导致错误。

示例 6.19：从 DBMS_SQL 包切换到 Native dynamic SQL。

功能描述：在本例中，应用 DBMS_SQL.TO_REFCURSOR 函数从 DBMS_SQL 包切换到 Native dynamic SQL。

程序代码如下。

```
DECLARE
    v_cur number;
    sql_string varchar2(1024);
    custid customers.custid%TYPE = 1;
    custname customers.custname%TYPE;
    gender customers.gender%TYPE;
    v_count int;
    cur refcursor;
BEGIN
```

```
   --打开游标
   v_cur := DBMS_SQL.open_cursor();

   --解析语句
   sql_string := 'select custname, gender from customers where custid = :
custid';
   DBMS_SQL.parse(v_cur, sql_string, DBMS_SQL.native);

   --绑定变量
   DBMS_SQL.bind_variable(v_cur, ':custid', custid);

   --执行语句
   v_count := DBMS_SQL.execute(v_cur);

   --将 DBMS_SQL 游标转换为 REF_CURSOR
   cur := DBMS_SQL.to_refcursor(v_cur);
   LOOP
     fetch cur into custname, gender;
     exit when cur%notfound;
     RAISE NOTICE 'custname: %, gender: %',custname,gender;
   END LOOP;
   close cur;
END;
```

程序运行结果如下。

```
NOTICE:custname: 林心水, gender: 男
```

4. DBMS_SQL.TO_CURSOR_NUMBER 函数

DBMS_SQL.TO_CURSOR_NUMBER 函数将 REFCURSOR 变量（强或弱）转换为 SQL 游标句柄，可以将其传递给 DBMS_SQL 子程序。将 REFCURSOR 变量传递给 DBMS_SQL.TO_CURSOR_NUMBER 函数之前，必须先打开它。将 REFCURSOR 变量转换为 SQL 游标句柄后，Native dynamic SQL 操作无法访问它。

示例 6.20：从 Native dynamic SQL 切换到 DBMS_SQL 包。

功能描述：本例应用 DBMS_SQL.TO_CURSOR_NUMBER 函数，从 Native dynamic SQL 切换到 DBMS_SQL 包。

程序代码如下。

```
DECLARE
    v_cur int;
    sql_string varchar2(1024);
    custname customers.custname%TYPE;
    gender customers.gender%TYPE;
    v_count int;
    cur refcursor;
BEGIN
```

```
    open cur for select custname, gender from customers limit 5;

    --将 REF_CURSOR 转换为 DBMS_SQL CURSOR
    v_cur := DBMS_SQL.TO_CURSOR_NUMBER(cur);

    DBMS_SQL.define_column(v_cur, 1, custname);
    DBMS_SQL.define_column(v_cur, 2, gender);

    LOOP
      v_count := DBMS_SQL.fetch_rows(v_cur);
      exit when v_count <= 0;
      DBMS_SQL.column_value(v_cur, 1, custname);
      DBMS_SQL.column_value(v_cur, 2, gender);
      RAISE NOTICE 'custname: %, gender: %',custname ,gender;
    END LOOP;

    DBMS_SQL.close_cursor(v_cur);
END;
```

程序运行结果如下。

```
NOTICE:custname: 林心水, gender: 男
NOTICE:custname: 江文曜, gender: 女
NOTICE:custname: 吕浩初, gender: 女
NOTICE:custname: 萧承望, gender: 女
NOTICE:custname: 程同光, gender: 女
```

6.4 SQL 注入

SQL 注入是恶意利用在 SQL 语句中使用客户端提供的数据的应用程序,从而获得对数据库未经授权的访问,以查看或操作受限数据。

本节介绍 PL/SQL 中的 SQL 注入漏洞,并解释如何防范。

6.4.1 SQL 注入技术

所有 SQL 注入技术都利用了一个漏洞,即字符串输入未正确验证并连接到动态 SQL 语句中。

1. 语句修改

语句修改意味着故意更改动态 SQL 语句,使其以程序开发人员意想不到的方式运行。

通常,用户通过更改 SELECT 语句的 WHERE 子句或插入 UNION ALL 子句来检索未经授权的数据。该技术的经典示例是通过使 WHERE 子句始终为 TRUE 来绕过密码验证。

示例 6.21:易受语句修改的过程。

功能描述:本例创建一个易受语句修改影响的存储过程,然后在修改和不修改语句的

情况下调用该存储过程。通过语句修改,该存储过程返回一个秘密记录。

程序代码如下。

```
CREATE OR REPLACE PROCEDURE get_dob (
  name   IN   customers.custname%TYPE,
  dob    OUT  customers.dob%TYPE
) AUTHID DEFINER
IS
  query VARCHAR2(4000);
BEGIN
  --后面的 SELECT 语句很容易被修改,因为它使用连接来构建 WHERE 子句。
  query := 'SELECT dob FROM customers WHERE custname='''
       || name
       || '''';
  RAISE NOTICE 'query: %', query;
    EXECUTE IMMEDIATE query INTO dob;
    RAISE NOTICE 'dob: %',dob;
END;
```

演示没有 SQL 注入的过程。

```
DECLARE
  dob customers.dob%TYPE;
BEGIN
  get_dob('林心水', dob);
END;
```

程序运行结果如下。

```
NOTICE:query: SELECT dob FROM customers WHERE custname='林心水'
NOTICE:dob: 1997-05-08 00:00:00
```

程序代码修改示例如下。

```
DECLARE
  dob customers.dob%TYPE;
BEGIN
  get_dob(
  'Anybody'' OR custname = ''江文曜''--',
  dob);
END;
```

程序运行结果如下。

```
NOTICE:query: SELECT dob FROM customers WHERE custname='Anybody' OR custname = '江文曜'--'
NOTICE:dob: 1996-01-30 00:00:00
```

2. 语句注入

语句注入是指用户将一条或多条 SQL 语句附加到动态 SQL 语句中。匿名 PL/SQL 块容易受到这种技术的攻击。

示例 6.22：易受语句注入攻击的存储过程。

功能描述：本例创建一个易受语句注入攻击的存储过程，然后在使用和不使用语句注入的情况下调用该存储过程。通过语句注入，该存储过程删除了本例中的记录。

程序代码如下。

```
CREATE OR REPLACE PROCEDURE p (
  id   IN   INT,
  NAME  IN  VARCHAR2
) AUTHID DEFINER
IS
  block VARCHAR2(4000);
BEGIN
  --以下块容易受到语句注入的影响，因为它是通过连接构建的
  block :=
    'BEGIN
      DBMS_OUTPUT.PUT_LINE(''id: ' || TO_CHAR(id) || ''');'
    || 'DBMS_OUTPUT.PUT_LINE(''name: ' || name || ''');'
    END;';

  DBMS_OUTPUT.PUT_LINE('block:' || block);

  EXECUTE IMMEDIATE block;
END;
```

没有 SQL 注入的操作的程序代码如下。

```
SET SERVEROUTPUT ON

BEGIN
  p(1, 'xx');
END;
```

程序运行结果如下。

```
block:BEGIN
  DBMS_OUTPUT.PUT_LINE('id: 1');DBMS_OUTPUT.PUT_LINE('name: xx');
  END;
id: 1
name: xx
```

查询程序代码如下。

```
INSERT INTO customers(custid,custname,gender)
values (10001, 'Test', '男');
SELECT custid,custname,gender FROM customers where custid>10000 ;
```

程序运行结果如下。

```
custid | custname | gender |
------+--------- +------+
10001  | Test     | 男    |
```

添加 SQL 注入的操作的程序代码如下。

```
SET SERVEROUTPUT ON
BEGIN
  p(1, 'Anything'');
  DELETE FROM customers WHERE custname=INITCAP(''Test');
END;
```

程序运行结果如下。

```
block:BEGIN
  DBMS_OUTPUT.PUT_LINE('id: 1');DBMS_OUTPUT.PUT_LINE('name: Anything');
DELETE FROM customers WHERE custname=INITCAP('Test');
    END;
id: 1
name: Anything
```

查询程序代码如下。

```
SELECT custid,custname,gender FROM customers where custid>10000 ;
```

程序运行结果如下。

```
custid | custname | gender |
------+--------- +------+
```

6.4.2 防范 SQL 注入

如果在 PL/SQL 应用程序中使用动态 SQL,必须检查输入文本,以确保它与预期完全一致。可以使用以下技术。

(1) 绑定变量。
(2) 验证检查。
(3) 显式格式模型。

使 PL/SQL 代码不受 SQL 注入攻击的最有效方法是使用绑定变量。数据库专门使用绑定变量的值,并且不以任何方式解释它们的内容。绑定变量也可以提高性能。

示例 6.23:绑定变量,以防止 SQL 注入。

功能描述:本例中的过程不受 SQL 注入攻击,因为它使用绑定变量构建动态 SQL 语句。

程序代码如下。

```
CREATE OR REPLACE PROCEDURE get_dob_2 (
  custid    IN  customers.custid%TYPE,
  custname  IN  customers.custname%TYPE,
  dob       OUT customers.dob%TYPE
) AUTHID DEFINER
IS
  query VARCHAR2(4000);
BEGIN
  query := 'SELECT dob FROM customers
      WHERE custid=:a AND custname=:b';

  RAISE NOTICE 'Query: %', query;

  EXECUTE IMMEDIATE query INTO dob USING custid, custname;
  RAISE NOTICE 'dob: %', dob;
END;
```

没有 SQL 注入的程序代码如下。

```
DECLARE
  dob customers.dob%TYPE;
BEGIN
  get_dob_2(1, '林心水', dob);
END;
```

程序运行结果如下。

```
NOTICE:Query: SELECT dob FROM customers
      WHERE custid=:a AND custname=:b
NOTICE:dob: 1997-05-08 00:00:00
```

程序代码修改如下。

```
DECLARE
  dob customers.dob%TYPE;
BEGIN
  get_dob_2(1,
      'Anybody'' OR name = ''zs''--',
      dob);
END;
```

程序运行结果如下。

```
NOTICE:Query: SELECT dob FROM customers
      WHERE custid=:a AND
      custname=:b
SQL 错误 [P0002]: 错误: 查询没有返回记录
```

```
Where: PL/SQL 函数 get_dob_2(integer,varchar,date)的第 10 行的 EXECUTE
SQL 语句 "CALL get_dob_2(1,
        'Anybody'' OR name = ''zs''--',
        dob)"
PL/SQL 函数 inline_code_block 的第 4 行的 CALL
```

第 7 章

异 常 处 理

在运行过程中,PL/SQL 程序难免出现各种错误,好的程序应该能处理异常,并保证程序正常运行。本章解释如何处理 PL/SQL 运行时的错误,即异常处理,主要介绍以下内容。
- 异常处理的概念和术语。
- 定义异常。
- 引发异常。
- 处理异常。

7.1 异常处理的概念和术语

异常(PL/SQL 程序运行时产生的错误)可能来自设计错误、编码错误、硬件故障或其他来源。无法预先处理所有可能触发的异常,但可以编写异常处理程序,让程序在触发异常的情况下还能继续运行。

任何 PL/SQL 块都可以有一个异常处理部分,它可以处理一个或多个异常。例如:

```
EXCEPTION
  WHEN ex_name_1 THEN statements_1                    --Exception handler
  WHEN ex_name_2 OR ex_name_3 THEN statements_2       --Exception handler
  WHEN OTHERS THEN statements_3                       --Exception handler
END;
```

在上面的语法示例中,ex_name_n(n 为 1,2,…)是异常的名称,statements_n(n 为 1,2,…)是一个或多个语句,还可以没有任何语句。

当 PL/SQL 块的可执行部分触发异常时,可执行部分会停止执行,并将控制权转移到异常处理部分。如果抛出异常 ex_name_1,则运行语句 statements_1;如果抛出异常 ex_name_2 或 ex_name_3,则运行语句 statements_2;如果抛出其他异常,则运行语句 statements_3。

异常处理程序运行后,控制权转移到 EXCEPTION 块的下一条语句。如果没有该块,则按以下原则执行。

(1) 如果异常处理程序在子程序中,则将控制权返回给调用者调用之后的语句处。

(2) 如果异常处理程序位于匿名块中,则将控制权转移到主机环境(如 KSQL)。

如果在没有异常处理程序的 PL/SQL 块中触发异常,则异常会传播。也就是说,异常会在连续的封闭块中向上抛出,直到一个 PL/SQL 块有一个异常处理程序或没有封闭块为止。如果没有异常处理程序,那么 PL/SQL 会向调用者或主机环境返回一个未处理的异常,这将决定最终的返回结果。

7.1.1 异常种类

异常的种类分为以下两种。

(1) 系统预定义异常。

系统预定义异常是 PL/SQL 已命名的异常,这些异常都有一个错误代码,且会在系统运行出错时隐式(自动)触发。

(2) 用户自定义异常。

可以在任何 PL/SQL 匿名块、子程序或包的声明部分中声明自己的异常。例如,可以声明一个名为 invalid_number 的异常,用于标记一个无效数字。

用户自定义异常必须显式触发。

系统预定义异常和用户自定义异常的差异如表 7.1 所示。

表 7.1 异常的差异

异 常 种 类	定 义	错误代码	名字	隐式触发	显示触发
系统预定义异常	系统	总有	总有	是	可选
用户自定义异常	用户	用户分配	总有	否	总是

对于命名异常,可以编写特定的异常处理程序,而不是使用 OTHERS 异常处理程序来处理它。特定的异常处理程序比 OTHERS 异常处理程序更有效,后者必须调用一个函数,确定它正在处理哪个异常。

7.1.2 异常处理程序的优点

使用异常处理程序进行异常处理将使程序更易于编写和理解,同时降低了未处理异常的可能性。

程序中如果没有异常处理部分,则必须检查所有可能触发的异常,并处理它。但也容易忽略可能出现的异常,尤其是在无法立即检测到异常的情况下(例如,使用了错误数据,在计算运行之前可能无法检测到)。异常处理代码可以分散在整个程序中。

使用异常处理程序,不需要预先知道每个可能触发的异常或它可能发生的位置,只需在可能发生错误的每个块中包含一个异常处理模块。在异常处理模块中,可以编写处理特定错误或未知错误的异常处理程序。如果块中的任何地方(包括子块内)发生错误,则异常处理程序都会捕获并处理它。错误处理的相关信息则会被隔离在块的异常处理部分,不再影响程序后续的执行。

在示例 7.1 中,一个存储过程使用单个异常处理程序来处理预定义异常 NO_DATA_FOUND,该异常可以出现在两个 SELECT INTO 语句中的任何一个。

如果多个语句使用相同的异常处理程序,并且想知道哪个语句触发了异常,可以使用变量辅助定位,如示例 7.2 所示。

如果确定了处理哪个异常,可以为特定异常设置一个异常处理程序,还可以通过将语句放入具有自己的异常处理程序块中检查单个语句中的异常。

示例 7.1:在单个异常处理程序中处理多个异常。

程序代码如下。

```
CREATE OR REPLACE PROCEDURE select_item (
  t_column VARCHAR2,
  t_name   VARCHAR2
) AUTHID DEFINER
IS
  temp VARCHAR2(30);
BEGIN
  temp := t_column;                    --For error message if next SELECT fails

  --Fails if table t_name does not have column t_column:
  SELECT COLUMN_NAME INTO temp
  FROM USER_TAB_COLS
  WHERE TABLE_NAME = UPPER(t_name)
  AND COLUMN_NAME = UPPER(t_column);

  temp := t_name;                      --For error message if next SELECT fails

  --Fails if there is no table named t_name:
  SELECT OBJECT_NAME INTO temp
  FROM USER_OBJECTS
  WHERE OBJECT_NAME = UPPER(t_name)
  AND OBJECT_TYPE = 'TABLE';

EXCEPTION
  WHEN NO_DATA_FOUND THEN
    RAISE NOTICE 'No Data found for SELECT on %',temp;
  WHEN OTHERS THEN
    RAISE NOTICE 'Unexpected error';
    RAISE;
END;
```

调用过程(有一个 SYS_PROC 表,但没有 NAME 列)的程序代码如下。

```
BEGIN
  select_item('sys_proc', 'name');
END;
```

程序运行结果如下。

```
No Data found for SELECT on sys_proc
```

调用过程(没有 EMP 表)的程序代码如下。

```
BEGIN
  select_item('emp', 'name');
END;
```

程序运行结果如下。

```
No Data found for SELECT on emp
```

示例 7.2：在共享异常处理程序中使用定位变量。

程序代码如下。

```
CREATE OR REPLACE PROCEDURE loc_var AUTHID DEFINER IS
  stmt_no   POSITIVE;
  name_   VARCHAR2(100);
BEGIN
  stmt_no := 1;

  SELECT table_name INTO name_
  FROM user_tables
  WHERE table_name LIKE 'SYS%' LIMIT 1;

  stmt_no := 2;

  SELECT table_name INTO name_
  FROM user_tables
  WHERE table_name LIKE 'XYZ%';
EXCEPTION
  WHEN NO_DATA_FOUND THEN
    RAISE NOTICE 'Table name not found in query  %',stmt_no;
END;

CALL loc_var();
```

程序运行结果如下。

```
Table name not found in query 2
```

7.2 定义异常

7.2.1 系统预定义异常

针对常见的异常以及系统运行时触发的异常，PL/SQL 内部为其预定义了一个名称。例如：除零错误，对应的预定义异常名称为 DIVISION_BY_ZERO。当错误发生时，系统隐式（自动）抛出该异常。PL/SQL 预定义异常如表 7.2 所示。

表 7.2　PL/SQL 预定义异常

KingbaseES 异常名称	Oracle 对应异常名称	异 常 说 明
CASE_NOT_FOUND	CASE_NOT_FOUND	CASE 语句中没有任何 WHEN 子句满足条件,且没有 ELSE 子句
COLLECTION_IS_NULL	COLLECTION_IS_NULL	调用一个未初始化的嵌套表或可变数组的方法(不包含 EXISTS),或为一个未初始化的嵌套表或可变数组的元素赋值
DUPLICATE_CURSOR	CURSOR_ALREADY_OPEN	打开一个已经打开的游标
UNIQUE_VIOLATION	DUP_VAL_ON_INDEX	给一个有唯一约束条件的数据字段保存相同的值
INVALID_CURSOR_NAME	INVALID_CURSOR	操作一个不合法的游标。例如关闭一个未打开的游标
INVALID_TEXT_REPRESENTATION	INVALID_NUMBER	出现运算、转换、截位或长度的约束错误
NO_DATA_FOUND	NO_DATA_FOUND	未获取数据
INTERNAL_ERROR	PROGRAM_ERROR	PL/SQL 内部错误
SELF_IS_NULL	SELF_IS_NULL	调用一个为空对象的 MEMBER 方法
OUT_OF_MEMORY	STORAGE_ERROR	内存溢出
SUBSCRIPT_BEYOND_COUNT	SUBSCRIPT_BEYOND_COUNT	调用嵌套表或可变数组时,使用的下标索引超出对应元素的总个数
SUBSCRIPT_OUTSIDE_LIMIT	SUBSCRIPT_OUTSIDE_LIMIT	调用嵌套表或可变数组时,使用的下标索引不在合法范围内,如(-1)
TOO_MANY_ROWS	TOO_MANY_ROWS	返回太多的结果行
NUMERIC_VALUE_OUT_OF_RANGE	VALUE_ERROR	数值类型超过定义域
DIVISION_BY_ZERO	ZERO_DIVIDE	除零错误

更多预定义异常信息可通过数据库系统函数 SYS_GET_PREDEFINED_EXCEPTION_DETAIL 查询,详见 SYS_GET_PREDEFINED_EXCEPTION_DETAIL 函数。

示例 7.3 计算了一家公司的市盈率。如果公司的收益为零,则除法操作会引发预定义的异常 DIVISION_BY_ZERO,并且块的可执行部分将控制权转移到异常处理部分。

示例 7.4 通过使用错误检查代码避免示例 7.3 出现的异常。

示例 7.3:匿名块中处理 DIVISION_BY_ZERO。

程序代码如下。

```
DECLARE
  stock_price  NUMBER := 9.73;
  net_earnings  NUMBER := 0;
```

```
    pe_ratio   NUMBER;
BEGIN
  pe_ratio := stock_price / net_earnings;  --raises ZERO_DIVIDE exception
  DBMS_OUTPUT.PUT_LINE('Price/earnings ratio = ' || pe_ratio);
EXCEPTION
  WHEN ZERO_DIVIDE THEN
    RAISE NOTICE 'Company had zero earnings.';
    pe_ratio := NULL;
END;
```

程序运行结果如下。

```
Company had zero earnings.
```

示例7.4：匿名块中避免 DIVISION_BY_ZERO。

程序代码如下。

```
DECLARE
  stock_price   NUMBER := 9.73;
  net_earnings  NUMBER := 0;
  pe_ratio   NUMBER;
BEGIN
  pe_ratio :=
    CASE net_earnings
      WHEN 0 THEN NULL
      ELSE stock_price / net_earnings
    END;
END;
```

7.2.2　用户自定义异常

可以在任何 PL/SQL 块、函数、存储过程或者包中声明一个异常。

语法格式如下。

```
exception_name EXCEPTION ;
```

用户自定义异常必须被显式触发，有关异常触发的详细信息详见"引发异常"。

用户自定义异常可以与一个错误码绑定，语法格式如下。

```
PRAGMA EXCEPTION_INIT (exception, error_code) ;
```

其中 exception 是用户自定义的异常，error_code 是大于-1000000 且小于 0 的整数，error_code 可以是系统预定义异常的错误码。

注意：EXCEPTION_INIT 仅可对当前声明块中声明的自定义异常进行错误码绑定。

当为包声明中声明的自定义异常进行错误码绑定时，无法使用包名对异常名称进行修饰。语法格式如下。

```
PRAGMA EXCEPTION_INIT (package.exception, error_code);
```

7.2.3 重新声明预定义的异常

建议不要重新声明系统预定义的异常,即声明一个用户定义的异常名称,该名称是系统预定义的异常名称(有关系统预定义异常名称的列表,请参见系统预定义异常)。

如果重新声明了系统预定义的异常,本地声明将覆盖 STANDARD 包中的全局声明。为全局声明异常编写的异常处理程序将无法处理它,除非使用包名 STANDARD 限定它的名称。

示例 7.5:重新声明预定义异常。

程序代码如下。

```
DROP TABLE IF EXISTS t CASCADE;
CREATE TABLE t (c NUMBER(2,1));
```

在下面的块中,INSERT 语句隐式引发了 VALUE_ERROR 异常,异常处理程序捕获并处理了这个异常。

程序代码如下。

```
DECLARE
  default_number NUMBER := 0;
BEGIN
  INSERT INTO t VALUES(TO_NUMBER('100.001', '9G999'));
EXCEPTION
  WHEN VALUE_ERROR THEN
    RAISE NOTICE 'Substituting default value for invalid number.';
    INSERT INTO t VALUES(default_number);
END;
```

程序运行结果如下。

```
Substituting default value for invalid number.
```

以下块重新声明了预定义异常 VALUE_ERROR。当 INSERT 语句隐式触发预定义异常 VALUE_ERROR 时,异常处理程序不会处理它。

程序代码如下。

```
DECLARE
  default_number NUMBER := 0;
  value_error EXCEPTION;              --redeclare predefined exception
BEGIN
  INSERT INTO t VALUES(TO_NUMBER('100.001', '9G999'));
EXCEPTION
  WHEN VALUE_ERROR THEN
    RAISE NOTICE 'Substituting default value for invalid number.';
    INSERT INTO t VALUES(default_number);
END;
```

程序运行结果如下。

```
ERROR:  numeric field overflow
DETAIL:  A field with precision 2, scale 1 must round to an absolute
value less than 10^1.
CONTEXT:   SQL statement "INSERT INTO t VALUES(TO_NUMBER('100.001', '9G999'))"
PL/SQL function inline_code_block line 5 at SQL statement
```

如果在异常处理程序中使用包名 STANDARD 限定异常名称，则上述块中的异常处理程序将处理预定义的异常 VALUE_ERROR。

程序代码如下。

```
DECLARE
  default_number NUMBER := 0;
  value_error EXCEPTION;               --redeclare predefined exception
BEGIN
  INSERT INTO t VALUES(TO_NUMBER('100.001', '9G999'));
EXCEPTION
  WHEN STANDARD.VALUE_ERROR THEN
    RAISE NOTICE 'Substituting default value for invalid number.';
    INSERT INTO t VALUES(default_number);
END;
```

程序运行结果如下。

```
Substituting default value for invalid number.
```

7.3　引发异常

7.3.1　显式触发异常

要显式触发异常，可以使用 RAISE 语句或存储过程 RAISE_APPLICATION_ERROR。

1. RAISE 语句

RAISE 语句可以显式地触发一个异常。在异常处理程序之外，RAISE 语句必须指定异常名称。如果在异常处理程序内部，且省略了异常名称，那么该 RAISE 语句将重新引发当前正在处理的异常。

语法格式如下。

```
RAISE [ exception ];
```

其中，exception 可以是已定义的用户自定义异常，也可以是系统预定义异常。

省略 exception 的 RAISE 子句仅可在异常处理模块中使用。

下面是 RAISE 语句的通常使用方法。

(1) 使用 RAISE 语句触发用户自定义的异常。

在示例 7.6 中,存储过程声明了 1 个名为 past_due 的异常,使用 RAISE 语句显式触发它,并使用异常处理程序加以处理。

示例 7.6:声明、触发和处理用户自定义的异常。

程序代码如下。

```
CREATE PROCEDURE account_status (
  due_date DATE,
  today    DATE)
AUTHID DEFINER IS
  past_due   EXCEPTION;               --declare exception
BEGIN
  IF due_date < today THEN
    RAISE past_due;                   --explicitly raise exception
  END IF;
EXCEPTION
  WHEN past_due THEN                  --handle exception
    RAISE NOTICE 'Account past due.';
END;
```

调用存储过程如下。

```
BEGIN
  account_status (TO_DATE('01-JUL-2010', 'DD-MON-YYYY'),
       TO_DATE('09-JUL-2010', 'DD-MON-YYYY'));
END;
```

程序运行结果如下。

```
NOTICE:: Account past due.
```

(2) 使用 RAISE 语句触发系统预定义的异常。

系统预定义异常通常由系统运行时隐式触发,但也可以使用 RAISE 语句显式触发它们。当一个预定义异常拥有对应的异常处理程序时,无论是显式触发还是隐式触发,都会触发异常处理程序对相应的异常进行处理。

在示例 7.7 中,存储过程不论显式或隐式触发预定义异常 VALUE_ERROR,异常处理程序始终都会处理它。

示例 7.7:触发系统预定义异常。

程序代码如下。

```
DROP TABLE IF EXISTS t CASCADE;
CREATE TABLE t (c NUMBER(2,1));

CREATE OR REPLACE PROCEDURE p (n NUMBER) AUTHID DEFINER IS
  default_number NUMBER := 0;
BEGIN
  IF n < 0 THEN
```

```
      RAISE VALUE_ERROR;                                         --显式触发
    ELSE
      INSERT INTO t VALUES(TO_NUMBER('100.001', '9G999'));       --隐式触发
    END IF;
EXCEPTION
  WHEN VALUE_ERROR THEN
    RAISE NOTICE 'Substituting default value for invalid number.';
    INSERT INTO t VALUES(default_number);
END;
```

调用存储过程的程序代码如下。

```
BEGIN
  p(-1);
END;
```

程序运行结果如下。

```
NOTICE:: Substituting default value for invalid number.
```

再次调用存储过程的程序代码如下。

```
BEGIN
  p(1);
END;
```

程序运行结果如下。

```
NOTICE:   Substituting default value for invalid number.
```

(3) 使用 RAISE 语句重新触发当前异常。

在异常处理程序中,可以使用 RAISE 语句重新引发当前正在处理的异常。重新引发的异常会将异常传递给当前块中的异常处理程序块。若当前块中无相应的异常处理块,则该异常会传播,详见 7.3.2 节"异常传播"。重新引发异常时,可以省略异常名称。

在示例 7.8 中,异常处理从内部块开始,到外部块结束。在外部块声明异常,因此异常对两个块都可见,并且每个块都有一个专门针对该异常的异常处理程序。在内部块触发异常,其异常处理程序进行初步处理,然后重新引发异常,将其传递给外部块进一步处理。

示例 7.8:重新触发异常。

程序代码如下。

```
DECLARE
    salary_too_high    EXCEPTION;
    current_salary   NUMBER := 20000;
    max_salary      NUMBER := 10000;
    erroneous_salary   NUMBER;
BEGIN
```

```
BEGIN
  IF current_salary > max_salary THEN
    RAISE salary_too_high;        --raise exception
  END IF;
EXCEPTION
  WHEN salary_too_high THEN       --start handling exception
    erroneous_salary := current_salary;
    RAISE NOTICE 'Salary % is out of range.', erroneous_salary;
    RAISE NOTICE 'Maximum salary is %.', max_salary;
    RAISE;          --reraise current exception (exception name is optional)
END;

EXCEPTION
  WHEN salary_too_high THEN       --finish handling exception
    current_salary := max_salary;

    RAISE NOTICE 'Revising salary from % to %.', erroneous_salary, current_salary;
END;
```

程序运行结果如下。

```
NOTICE:  Salary 20000 is out of range.
NOTICE:  Maximum salary is 10000.
NOTICE:  Revising salary from 20000 to 10000.
```

2. 存储过程 RAISE_APPLICATION_ERROR

存储过程 RAISE_APPLICATION_ERROR 通常用来抛出一个用户自定义异常，并将错误码和错误信息传送给调用者。

调用 RAISE_APPLICATION_ERROR 的语法格式如下。

```
RAISE_APPLICATION_ERROR (error_code, message);
```

其中 error_code 是 −20999～−20000 的整数，message 是长度为 2048B 的字符串，大于该长度的字符会被自动截断。

使用 RAISE_APPLICATION_ERROR 抛出的用户自定义异常必须使用 PRAGMA EXCEPTION_INIT 分配的 error_code。

通过存储过程 RAISE_APPLICATION_ERROR 抛出的异常可以通过 OTHERS 捕获处理，也可以通过绑定了相同 error_code 的用户自定义异常加以捕获处理，详见用户自定义异常。

在示例 7.9 中，一个匿名块中声明了一个名为 past_due 的异常，并为其分配了错误代码-20000。然后调用一个存储过程，存储过程调用带有错误代码-20000 和消息的存储过程 RAISE_APPLICATION_ERROR，之后控制权返回匿名块异常处理部分。为了检索与异常关联的消息，匿名块中的异常处理程序调用 SQLERRM 函数，详见"检索异常信息"。

示例 7.9：使用 RAISE_APPLICATION_ERROR 引发用户自定义的异常。

程序代码如下。

```
CREATE OR REPLACE PROCEDURE account_status (
  due_date DATE,
  today    DATE
) AUTHID DEFINER IS
BEGIN
  IF due_date < today THEN            --explicitly raise exception
    RAISE_APPLICATION_ERROR(-20000, 'Account past due.');
  END IF;
END;
```

调用存储过程的代码如下。

```
DECLARE
  past_due   EXCEPTION;               --declare exception
  PRAGMA EXCEPTION_INIT (past_due, -20000); -- assign error code to exception
BEGIN
  account_status (TO_DATE('01-JUL-2010', 'DD-MON-YYYY'),
       TO_DATE('09-JUL-2010', 'DD-MON-YYYY')); -- invoke procedure

EXCEPTION
  WHEN past_due THEN                  --handle exception
    RAISE NOTICE '%', SQLERRM(-20000);
END;
```

程序运行结果如下。

```
NOTICE:  ERRCODE-20000: Account past due.
```

7.3.2 异常传播

如果在一个没有异常处理程序的块中抛出异常,则该异常会传播,即该异常会在连续的封闭块中重现,直到被某个具有对应异常处理程序的块捕获处理或者没有封闭块为止。若一直没有对应的异常处理程序,那么 PL/SQL 会将未处理的异常返回调用者或返回到主机环境中,该异常也会成为当前程序执行的结果。

用户自定义的异常可以传播到超出其声明的范围(即超出声明它的块)。但由于其名称在超出的范围内不存在,因此,超出范围后,用户自定义的异常只能使用 OTHERS 来捕获处理。

在示例 7.10 中,内部块声明了一个名为 past_due 的异常,它没有异常处理程序。当内部块引发 past_due 时,异常传播到外部块,其中不存在名称 past_due。外部块使用 OTHERS 异常处理程序处理异常,如果外部块不处理用户自定义的异常,则会发生错误,如示例 7.11 所示。

示例7.10：处理超出范围的传播异常。

程序代码如下。

```
CREATE OR REPLACE PROCEDURE p AUTHID DEFINER AS
BEGIN

  DECLARE
    past_due   EXCEPTION;
    due_date   DATE := trunc(SYSDATE) - 1;
    todays_date  DATE := trunc(SYSDATE);
  BEGIN
    IF due_date < todays_date THEN
      RAISE past_due;
    END IF;
  END;

EXCEPTION
  WHEN OTHERS THEN
    NULL;
END;
```

示例7.11：未处理超出范围的传播异常。

程序代码如下。

```
BEGIN
  DECLARE
    past_due   EXCEPTION;
    due_date   DATE := trunc(SYSDATE) - 1;
    todays_date  DATE := trunc(SYSDATE);
  BEGIN
    IF due_date < todays_date THEN
      RAISE past_due;
    END IF;
  END;
END;
```

程序运行结果如下。

```
ERROR:  unhandled exception "past_due"
CONTEXT:  PL/SQL function inline_code_block line 8 at RAISE
```

1. 未传播的异常

异常在块中被触发并捕获处理，未传播，如示例7.12所示。

示例7.12：未传播的异常。

程序代码如下。

```
CREATE OR REPLACE PROCEDURE p AS
BEGIN
```

```
DECLARE
  past_due   EXCEPTION;
BEGIN
  RAISE past_due;
EXCEPTION
  WHEN past_due THEN
    RAISE NOTICE 'Exception Does Not Propagate';
  END;
END;
```

调用存储过程的代码如下。

```
CALL P();
```

程序运行结果如下。

```
NOTICE:  Exception Does Not Propagate
```

2. 从内部块传播到外部块的异常

异常在内部块中被触发,但未被捕获,传播到上层块中被捕获处理,如示例 7.13 所示。

示例 7.13:从内部块传播到外部块的异常。

程序代码如下。

```
CREATE OR REPLACE PROCEDURE p AS
BEGIN
  DECLARE
    past_due   EXCEPTION;
  BEGIN
    RAISE past_due;
  END;
EXCEPTION
  WHEN OTHERS THEN
    RAISE NOTICE 'Exception Propagates from Inner Block to Outer Block';
END;
```

调用存储过程的代码如下。

```
CALL P();
```

程序运行结果如下。

```
NOTICE:  Exception Propagates from Inner Block to Outer Block
```

3. 声明中触发的异常的传播

当前块中声明阶段抛出的异常将直接传播到上层块或主机环境,而非当前块。因此,声明阶段的异常处理程序必须限定在调用块中,而不是在声明的当前块中。

在示例 7.14 中，VALUE_ERROR 异常处理程序与引发 VALUE_ERROR 的声明位于同一块中。因为异常会立即传播到主机环境中，所以异常处理程序不会处理它。

示例 7.15 与 7.14 类似，只是上层块中处理了内部块声明中引发的 VALUE_ERROR 异常。

示例 7.14：声明中触发的异常未处理。

程序代码如下。

```
DECLARE
  credit_limit CONSTANT NUMBER(3) := 5000;
BEGIN
  NULL;
EXCEPTION
  WHEN VALUE_ERROR THEN
    RAISE NOTICE 'Exception raised in declaration.';
END;
```

程序运行结果如下。

```
ERROR:  numeric field overflow
DETAIL:  A field with precision 3, scale 0 must round to an absolute value less than 10^3.
CONTEXT:  PL/SQL function inline_code_block line 3 during statement block local variable initialization
```

示例 7.15：声明中触发的异常由上层块处理。

程序代码如下。

```
BEGIN
  DECLARE
    credit_limit CONSTANT NUMBER(3) := 5000;
  BEGIN
    NULL;
  END;
EXCEPTION
  WHEN VALUE_ERROR THEN
    RAISE NOTICE 'Exception raised in declaration.';
END;
```

程序运行结果如下。

```
NOTICE:  Exception raised in declaration.
```

4. 异常处理程序中抛出的异常传播

异常处理阶段抛出的异常与声明阶段抛出的异常一样，会直接传播到上层块或主机环境中。因此，异常处理程序必须位于上层块或调用块中。

在示例 7.16 中，当 n 为 0 时，计算 $1/n$ 会引发预定义的异常 ZERO_DIVIDE，并且控制权转移到同一块中的 ZERO_DIVIDE 异常处理程序。当异常处理程序触发 ZERO_DIVIDE 时，异常会立即传播到调用程序。调用者不处理异常，所以 PL/SQL 向主机环境返回一个未处理的异常错误。

示例 7.17 与 7.16 类似，只是当存储过程向调用者返回未处理的异常错误时，调用者会处理它。

示例 7.18 与 7.17 类似，只是由上层块处理内部块中的异常处理程序引发的异常。

示例 7.19 中，存储过程中的异常处理部分具有用于处理用户自定义异常 i_is_one 和预定义异常 ZERO_DIVIDE 的异常处理程序。当 i_is_one 异常处理程序引发 ZERO_DIVIDE 时，异常会立即传播到调用程序中（ZERO_DIVIDE 异常处理程序不会处理它）。调用者不处理异常，所以 PL/SQL 向主机环境返回一个未处理的异常错误。

示例 7.20 与 7.19 类似，只是上层的块处理了 i_is_one 异常处理程序引发的 ZERO_DIVIDE 异常。

示例 7.16：异常处理程序中触发的异常未处理。

程序代码如下。

```
CREATE OR REPLACE PROCEDURE print_reciprocal (n NUMBER) AUTHID DEFINER IS
BEGIN
  RAISE NOTICE '%', 1/n;
EXCEPTION
  WHEN ZERO_DIVIDE THEN
    RAISE NOTICE 'Error:';
    RAISE NOTICE '% is undefined', 1/n;
END;
```

调用存储过程的代码如下。

```
BEGIN  --invoking block
  print_reciprocal(0);
END;
```

程序运行结果如下。

```
NOTICE:  Error:
ERROR:  division by zero
CONTEXT:  PL/SQL function print_reciprocal(numeric) line 7 at RAISE
SQL statement "CALL print_reciprocal(0)"
PL/SQL function inline_code_block line 2 at CALL
```

示例 7.17：异常处理程序中引发的异常由调用程序处理。

程序代码如下。

```
CREATE OR REPLACE PROCEDURE print_reciprocal (n NUMBER) AUTHID DEFINER IS
BEGIN
  RAISE NOTICE '%', 1/n;
```

```
EXCEPTION
  WHEN ZERO_DIVIDE THEN
    RAISE NOTICE 'Error:';
    RAISE NOTICE '% is undefined', 1/n;
END;
```

调用存储过程的代码如下。

```
BEGIN                         --invoking block
  print_reciprocal(0);
EXCEPTION
  WHEN ZERO_DIVIDE THEN        --handles exception raised in exception handler
    RAISE NOTICE '1/0 is undefined.';
END;
```

程序运行结果如下。

```
NOTICE:  Error:
NOTICE:  1/0 is undefined.
```

示例 7.18：异常处理程序中引发的异常由上层的块处理。

程序代码如下。

```
CREATE OR REPLACE PROCEDURE print_reciprocal (n NUMBER) AUTHID DEFINER IS
BEGIN
  BEGIN
    RAISE NOTICE '%',1/n;
  EXCEPTION
    WHEN ZERO_DIVIDE THEN
      RAISE NOTICE 'Error in inner block:';
      RAISE NOTICE '1/n is undefined.';
  END;

EXCEPTION
  WHEN ZERO_DIVIDE THEN        --handles exception raised in exception handler
    RAISE NOTICE 'Error in outer block: ';
    RAISE NOTICE '1/0 is undefined.';
END;
```

调用存储过程的代码如下。

```
BEGIN
  print_reciprocal(0);
END;
```

程序运行结果如下。

```
Error in inner block:
Error in outer block: 1/0 is undefined.
```

示例 7.19：未处理异常处理程序中引发的异常。

程序代码如下。

```
CREATE OR REPLACE PROCEDURE descending_reciprocals (n INTEGER) AUTHID DEFINER IS
  i INTEGER;
  i_is_one EXCEPTION;
BEGIN
  i := n;

  LOOP
    IF i = 1 THEN
      RAISE i_is_one;
    ELSE
      RAISE NOTICE 'Reciprocal of % is %', i, 1/i;
    END IF;

    i := i - 1;
  END LOOP;
EXCEPTION
  WHEN i_is_one THEN
    RAISE NOTICE '1 is its own reciprocal.';
    RAISE NOTICE 'Reciprocal of % is %',TO_CHAR(i-1),TO_CHAR(1/(i-1));

  WHEN ZERO_DIVIDE THEN
    RAISE NOTICE 'Error:';
    RAISE NOTICE '% is undefined',1/n;
END;
```

调用存储过程的代码如下。

```
BEGIN
  descending_reciprocals(3);
END;
```

程序运行结果如下。

```
ERROR:  division by zero
PL/SQL function descending_reciprocals(integer) line 17 at CALL
SQL statement "CALL descending_reciprocals(3)"
PL/SQL function inline_code_block line 2 at CALL
Reciprocal of 3 is 0
Reciprocal of 2 is 0
1 is its own reciprocal.
```

示例 7.20：异常处理程序中引发的异常由上层的块处理。

程序代码如下。

```
CREATE OR REPLACE PROCEDURE descending_reciprocals (n INTEGER) AUTHID DEFINER IS
  i INTEGER;
```

```
    i_is_one EXCEPTION;
BEGIN
  BEGIN
    i := n;

    LOOP
      IF i = 1 THEN
        RAISE i_is_one;
      ELSE
        RAISE NOTICE 'Reciprocal of % is %',i,(1/i);
      END IF;

      i := i - 1;
    END LOOP;
  EXCEPTION
    WHEN i_is_one THEN
      RAISE NOTICE '1 is its own reciprocal.';
      RAISE NOTICE 'Reciprocal of % is %',TO_CHAR(i-1),TO_CHAR(1/(i-1));

    WHEN ZERO_DIVIDE THEN
      RAISE NOTICE 'Error:';
      RAISE NOTICE '% is undefined',(1/n);
  END;

EXCEPTION
  WHEN ZERO_DIVIDE THEN   --handles exception raised in exception handler
    RAISE NOTICE 'Error:';
    RAISE NOTICE '1/0 is undefined';
END;
```

调用存储过程的代码如下。

```
BEGIN
  descending_reciprocals(3);
END;
```

程序运行结果如下。

```
Reciprocal of 3 is 0
Reciprocal of 2 is 0
1 is its own reciprocal.
Error:
1/0 is undefined
```

7.3.3 未处理的异常

异常在内部块中被触发,但未被捕获,且传播到外层块中仍未被捕获,最终返回到主机环境中。

在打开单语句回滚的情况下,如果异常未被处理,PL/SQL 则会发生事务级回滚。

提示:通过在每个 PL/SQL 程序的顶层包含一个 OTHERS 异常处理程序,可以避免未处理异常。

示例 **7.21**:未处理异常。

程序代码如下。

```
CREATE OR REPLACE PROCEDURE p AS
BEGIN
  DECLARE
    past_due   EXCEPTION;
    PRAGMA EXCEPTION_INIT (past_due, -4910);
  BEGIN
    RAISE past_due;
  END;
END;
```

调用存储过程的代码如下。

```
CALL P();
```

程序运行结果如下。

```
ERROR:  ERRCODE-4910: non-KingbaseES exception
CONTEXT:  PL/SQL function p() line 7 at RAISE
```

 ## 7.4 处理异常

7.4.1 处理异常的措施

为了使程序尽可能可靠和安全,处理异常的措施通常包括以下内容。

(1) 使用错误检索代码和异常处理程序。

在输入错误的数据可能导致出错的地方使用错误检查代码。比如,不正确或为空的参数和不返回行或返回的行数超出预期的查询语句。可以在程序编写过程中输入错误数据的不同组合来测试代码,以查看可能出现的错误。还可以使用错误检查代码避免引发异常。

(2) 在可能发生错误的地方添加异常处理程序。

在数字计算、字符串操作和数据库操作期间尤其容易出错。错误也可能来自与代码无关的问题,例如,磁盘存储或内存硬件故障,但代码仍然需要异常处理程序来处理它们。

(3) 开发人员设计的程序在数据库未处于所期望的状态时继续工作。

例如,所查询的表可能添加或删除了列,或者它们的类型已经更改。可以使用%TYPE限定符声明变量,或者使用%ROWTYPE限定符声明记录变量,保存查询结果,以避免出错。

(4) 尽可能使用命名异常编写异常处理程序,而不是使用 OTHERS 异常处理程序。

了解预定义异常的名称和原因。如果已经知道进行的数据库操作可能会引发特定的系统预定义异常,请专门为它们编写异常处理程序。

(5) 通过异常处理程序输出调试信息。

如果将调试信息存储在单独的表中,请使用自治事务程序执行此操作。这样,即使回滚主程序所做的工作,也可以提交调试信息。有关自治事务程序的信息,请参阅 5.3 节"自治事务"。

(6) 对于每个异常处理程序,仔细决定是让它进行事务的提交、回滚还是继续执行。

无论错误的严重程度如何,都需要使数据库保持正常状态,并避免存储错误数据。

(7) 通过在每个 PL/SQL 程序的顶层包含一个 OTHERS 异常处理程序来避免漏处理某些异常。

将 OTHERS 异常处理程序中的最后一条语句设为 RAISE,有关 RAISE 或调用 RAISE_APPLICATION_ERROR 的信息,请参阅 7.3.1 节"显式触发异常"。

7.4.2 检索异常信息

在异常处理程序中,对于正在处理的异常,可以使用 SQLCODE 函数中描述的 PL/SQL 函数 SQLCODE 检索错误代码。

也可以使用以下任一方法检索错误消息。

(1) PL/SQL 函数 SQLERRM,详见本节 SQLERRM 函数。

该函数最多返回 512B,这是数据库错误消息的最大长度(包括错误代码、详细的错误信息等)。

(2) PL/SQL 系统包 DBMS_UTILITY 中的包函数。DBMS_UTILITY.FORMAT_ERROR_STACK,此函数返回完整的错误堆栈,最多 2000B。

(3) KingbaseES 数据库提供的其他异常信息检索方式。

另请参阅:

PL/SQL 系统包中有关 DBMS_UTILITY.FORMAT_ERROR_BACKTRACE 函数的信息,该函数返回引发异常时的调用堆栈,即使该程序是从上层块的异常处理程序中被调用。

PL/SQL 系统包中有关 UTL_CALL_STACK 包的信息,其子程序提供有关当前正在执行的子程序信息,包括子程序名称。

可以通过以下变量或函数检索异常信息。

1. SQLSTATE 变量

异常发生时,调用 SQLSTATE 变量返回当前异常对应的 SQL 标准错误码(5 位字符串)。

2. SQLERRM 变量

异常发生时,调用 SQLERRM 变量返回当前异常相应的错误信息。

注意:变量 SQLSTATE 和 SQLERRM 仅在异常处理模块中可用。

3. SQLCODE 函数

语法格式如下。

```
INT SQLCODE ();
```

说明如下。

(1) 检索异常兼容 ORACLE 的错误码(负整数)。在异常处理程序之外,SQLCODE 函数始终返回 0,表示程序正常编译执行。

(2) 对于系统预定义的异常,错误码是关联相关异常错误的编号,该数字通常为一个负整数(NO_DATA_FOUND 除外,其错误码为+100)。

(3) 对于用户自定义的异常,数字代码为+1(默认值),或者通过 EXCEPTION_INIT 方法绑定的错误码。

4. SQLERRM 函数

语法格式如下。

```
TEXT SQLERRM ( [ error_code ] ) ;
```

说明如下。

(1) 检索异常错误信息。

(2) 其中 error_code 为数据库异常对应的错误码,省略该参数时,返回当前正在处理的异常错误信息。当 error_code 不是一个标准的错误码或者不存在时,都会进行相应的错误提示。

5. ERROR_LINE 函数

语法格式如下。

```
INT ERROR_LINE () ;
```

说明如下。

返回抛出异常的函数(存储过程)中抛出异常的位置行号,若没有在异常处理块中,则返回 NULL。

6. ERROR_PROCEDURE 函数

语法格式如下。

```
VARCHAR(128) ERROR_PROCEDURE () ;
```

说明如下。

返回抛出异常的函数(存储过程)或触发器的名称,若没有在异常处理块中,则返回 NULL。

7. ERROR_NUMBER 函数

语法格式如下。

```
INT ERROR_NUMBER () ;
```

说明如下。

返回抛出异常的错误号,与 SQLCODE 相同。若没有在异常处理块中,则返回 NULL。

8. ERROR_MESSAGE 函数

语法格式如下。

```
VARCHAR(4000) ERROR_MESSAGE ();
```

说明如下。

返回异常消息的完整文本,与 SQLERRM 相同。若没有在异常处理块中,则返回 NULL。

9. ERROR_STATE 函数

语法格式如下。

```
INT ERROR_STATE ();
```

说明如下:

返回异常状态号,在异常处理块中始终为 1。若没有在异常处理块中,则返回 NULL。

10. ERROR_SEVERITY 函数

语法格式如下。

```
INT ERROR_SEVERITY ();
```

说明如下。

返回异常的严重级别,在异常处理块中始终为 20。若没有在异常处理块中,则返回 NULL。

11. SYS_GET_PREDEFINED_EXCEPTION_DETAIL 函数

语法格式如下。

```
SETOF RECORD SYS_GET_PREDEFINED_EXCEPTION_DETAIL (
    OUT exception_name TEXT,
    OUT sqlstate       TEXT,
    OUT sqlcode        INTEGER);
```

说明如下。

返回所有预定义异常的相关信息,exception_name 为预定义异常名称,sqlstate 为预定义异常的 SQL 标准错误码,sqlcode 为预定义异常的 ORACLE 兼容错误码,如示例 7.23 所示。

示例 7.22:SQLCODE 函数和 SQLERRM 函数。

使用错误检查函数避免引发异常,程序代码如下。

```
CREATE OR REPLACE PROCEDURE p AUTHID DEFINER AS
  i INT;
  v_code  NUMBER;
  v_errm  VARCHAR2(64);
```

```
BEGIN
  i := 1/0;
EXCEPTION
  WHEN OTHERS THEN
    v_code := SQLCODE;
    v_errm := SUBSTR(SQLERRM, 1, 64);
    RAISE NOTICE 'Error code %:%',v_code,v_errm;
END;
```

调用存储过程代码如下。

```
CALL P();
```

程序运行结果如下。

```
Error code -1476: division by zero
```

示例 7.23：SYS_GET_PREDEFINED_EXCEPTION_DETAIL 函数。

程序代码如下。

```
SELECT * FROM SYS_GET_PREDEFINED_EXCEPTION_DETAIL();
```

程序运行结果如下。

```
           exception_name            | sqlstate | sqlcode |
-------------------------------------+----------+---------+
 sql_statement_not_yet_complete      | 03000    | -10000  |
 connection_exception                | 08000    | -10001  |
 connection_does_not_exist           | 08003    | -10002  |
 ...
 tablespace_is_readonly              | F1001    | -10243  |
 tablespace_is_write                 | F1002    | -10244  |
 tablespace_operation_notsupport     | F1003    | -10245  |
```

7.4.3 异常捕获

异常捕获是异常处理程序的一部分，通常通过 EXCEPTION 子句对异常进行捕获操作。

通常情况下，一个语句引发异常将导致当前语句退出执行，与该语句处于同一个事务内的语句也会回滚。为了处理异常，PL/SQL 块中通过 EXCEPTION 子句捕获异常，进行相应处理，这时，处于同一个事务块中已经执行的语句不会回滚，但当前语句回滚。

在 KingbaseES 的 PL/SQL 中，当一条语句执行后，控制权将移交到下一条语句，但是当异常触发后，KingbaseES 的 PL/SQL 将立即捕获、处理异常。

语法格式如下。

```
[ DECLARE
```

```
    [ < VariableDeclaration > ]
    [ < CursorDeclaration > ]
    [ < UserDefinedExceptionDeclaration > ]
  ]
BEGIN
  < Statements >
EXCEPTION
  WHEN ExceptionName [ OR ExceptionName... ] THEN
    < HandlerStatements >;
  [ WHEN ExceptionName[ OR ExceptionName... ] THEN
    < HandlerStatements >;
  ... ]
END;
```

说明如下。

（1）ExceptionName 是异常的名称，它由系统预先定义或用户自定义，直接使用即可，如 division_by_zero，表示发生"除零"错误。异常名与大小写无关，一个异常名也可以通过 SQLSTATE 代码指定，例如，以下代码是等价的。

```
WHEN division_by_zero THEN ...
WHEN SQLSTATE '22012' THEN ...
```

（2）ExceptionName 可以使用关键字 OTHERS，用于处理在 OTHERS 之前没有显示指定处理的异常。

（3）如果没有发生异常，这种形式的块只是简单地执行所有 Statements，然后转到 END 之后的下一个语句。但是如果 Statements 内发生了一个错误，则会放弃对 Statements 的进一步处理，然后转到 EXCEPTION 子句。系统会在异常条件列表中匹配当前触发的异常。如果匹配成功，则执行对应的 HandlerStatements，完成后转到 END 之后的下一个语句。如果匹配失败，该异常就会传播出去，就像没有 EXCEPTION 子句一样。异常可以被外层闭合块中的 EXCEPTION 子句捕获，如果没有 EXCEPTION，则中止该程序的执行。

（4）如果在选中的 HandlerStatements 内触发新的异常，那么它不能被当前这个 EXCEPTION 子句捕获，而是被传播出去。由外层的 EXCEPTION 子句捕获它。

（5）当一个异常被 EXCEPTION 子句捕获时，PL/SQL 函数的局部变量会保持异常触发时的值，但是该块中所有对数据库状态的改变都会回滚。如示例 7.24 所示。

（6）如果 KES 数据库打开了单语句回滚参数（ora_statement_level_rollback），触发异常的语句不会回滚之前已经执行完成的操作。如示例 7.25 所示。

（7）进入和退出一个包含 EXCEPTION 子句的块比不包含 EXCEPTION 的块开销大得多。因此，尽量在必要时使用 EXCEPTION 子句。

示例 7.24：异常触发导致 PL/SQL 函数内已完成的操作回滚。

程序代码如下。

```
DROP TABLE IF EXISTS mytab CASCADE;
CREATE TABLE mytab(firstname TEXT, lastname TEXT);
```

```
INSERT INTO mytab(firstname, lastname) VALUES('Tom', 'Jones');

CREATE OR REPLACE PROCEDURE PROC() AS
    x int := 1;
BEGIN
    UPDATE mytab SET firstname = 'Joe' WHERE lastname = 'Jones';
    x := x / 0;
EXCEPTION
    WHEN division_by_zero THEN
        RAISE NOTICE 'caught division_by_zero';
END;
```

调用函数的代码如下。

```
CALL PROC();
NOTICE:  caught division_by_zero
```

查询更改的代码如下。

```
SELECT * FROM mytab;

firstname | lastname |
----------+----------+
Tom       | Jones    |
(1 row)
```

当函数执行到对 y 赋值的位置时，会触发 division_by_zero 异常，该异常将被 EXCEPTION 子句捕获。而在 RETURN 语句中返回的值将是 x 执行加法后的值。不过，在该块之前的 INSERT 将不会回滚，因此，最终的结果是数据库包含"Tom Jones"，但不包含"Joe Jones"。

示例 7.25：异常触发未影响 PL/SQL 函数内已完成的操作。

程序代码如下。

```
set ora_statement_level_rollback to on;

DROP TABLE IF EXISTS mytab CASCADE;
CREATE TABLE mytab(firstname TEXT, lastname TEXT);
INSERT INTO mytab(firstname, lastname) VALUES('Tom', 'Jones');

CREATE OR REPLACE PROCEDURE PROC() AS
    x int := 1;
BEGIN
    UPDATE mytab SET firstname = 'Joe' WHERE lastname = 'Jones';
    x := x / 0;
EXCEPTION
    WHEN division_by_zero THEN
        RAISE NOTICE 'caught division_by_zero';
END;
```

调用函数的代码如下。

```
CALL PROC();
NOTICE:   caught division_by_zero
```

查询更改的代码如下。

```
SELECT * FROM mytab;
firstname | lastname |
----------+----------+
Joe       | Jones    |
(1 row)
```

7.4.4 获取异常状态信息

在异常处理模块中，除系统变量和系统函数外，KingbaseES 也支持使用 GET STACKED DIAGNOSTICS 命令获取有关当前异常的状态信息，该命令的语法格式如下。

```
GET STACKED DIAGNOSTICS variable { = | := } item [, ...];
```

每个 item 是一个关键字，标识一个被赋予到指定变量的状态值，可用的状态项如表 7.3 所示。

表 7.3 异常状态项

名　　称	类型	描　　述
RETURNED_SQLSTATE	text	该异常的 SQLSTATE 错误代码
COLUMN_NAME	text	与异常相关的列名
CONSTRAINT_NAME	text	与异常相关的约束名
MESSAGE_TEXT	text	该异常主要消息的文本
TABLE_NAME	text	与异常相关的表名
SCHEMA_NAME	text	与异常相关的模式名

对出现的异常信息，还需要获取相对应的执行位置信息，有利于找到问题根源。相关内容请参考第 2 章 2.6 节。

示例 7.26：获取 PL/SQL 中异常时的错误信息。

程序代码如下。

```
DECLARE
  text_var1 text;
  text_var2 text;
  text_var3 text;
BEGIN
  INSERT INTO categories(catgid,catgname,parentid)
  VALUES (100,'手机',10);
```

```
    EXCEPTION WHEN OTHERS THEN
      GET STACKED DIAGNOSTICS
        text_var1 = COLUMN_NAME,
        text_var2 = CONSTRAINT_NAME,
        text_var3 = MESSAGE_TEXT;
      RAISE NOTICE '%,%,%',text_var1,text_var2,text_var3;
    END;
```

程序运行结果如下。

```
NOTICE:,categories_pkey,重复键违反唯一约束"categories_pkey"
```

7.4.5 检查断言

ASSERT 语句是一种在 PL/SQL 函数中用于调试的语句，用来检查运行时的一些条件是否满足。

语法格式如下。

```
ASSERT condition [ , message ];
```

说明如下。

（1）condition 是一个布尔表达式，它的计算结果被断言为真。若结果为真，ASSERT 语句不会对程序产生任何影响；若结果为假或空，则将发生 ASSERT_FAILURE 异常（如果计算 condition 时发生错误，会抛出对应的普通异常）。

（2）如果 message 参数非空，则在断言失败时，message 会被用来替换默认的错误消息文本（"assertion failed"）。message 表达式在断言成功时不会对程序产生任何影响。

（3）通过配置参数 PL/SQL.check_asserts 可以启用或者禁用断言测试，这个参数为布尔值且默认为 on。如果参数值被设置为 off，则 ASSERT 语句什么也不做。

（4）ASSERT 是为了检测程序的 bug，而不是报告普通的错误情况。如果要报告普通错误，请使用前面介绍的打印语句。

示例 7.27：使用 assert 语句检查是否有姓氏为王的顾客。

程序代码如下。

```
DECLARE
    customer_count integer;
BEGIN
    SELECT count(*)
    INTO customer_count
    FROM customers
    WHERE custname LIKE '王%';
    assert customer_count > 0, '没有找到姓王的顾客';
END;
```

因为存在姓王的顾客记录，所以该语句块没有发出任何消息。

以下示例则发出错误消息，原因在于，对姓王的顾客人数统计没有超过100。

示例7.28：使用assert语句检查是否有姓氏为王的顾客人数超过100。

程序代码如下。

```
DECLARE
    customer_count integer;
BEGIN
    SELECT count(*)
    INTO customer_count
    FROM customers
    WHERE custname LIKE '王%';
    assert customer_count > 100, '姓王的顾客人数没有超过100';
END;
```

程序运行结果如下。

```
SQL 错误 [P0004]: 错误: 姓王的顾客人数没有超过100
Where: PL/SQL 函数 inline_code_block 的第 8 行的 ASSERT
```

第 8 章

PL/SQL 中的输入与输出

日常使用及开发的大部分 PL/SQL 程序,通过 SQL 和数据库服务交互足以满足要求。但是还有这样一些应用场景,例如,通过 PL/SQL 向外部环境发送信息,或者从一些外部数据源(如控制台、文件、网络等)读取信息等。本章将探讨 PL/SQL 中一些最常用的 I/O 机制,以解决上述问题。本章包括以下内容。

- DBMS_OUTPUT(用于在控制台输入和输出信息)。
- UTL_FILE(用于读取数据到内存和写入数据到本地)。
- UTL_HTTP(通过网络获取数据)。

8.1 显示信息(DBMS_OUTPUT)

系统提供 DBMS_OUTPUT 包,利用 DBMS_OUTPUT.PUT 或 DBMS_OUTPUT.PUTLINE 方法将数据传入缓存中,方便其他 PL/SQL 程序通过使用 DBMS_OUTPUT.GET_LINE 或 DBMS_OUTPUT.GET_LINES 获取数据。初始时,每一个用户的会话都具有一个默认大小为 20000B 的 DBMS_OUTPUT 缓存区,开发人员一般将其设置为 UNLIMITED。因为一旦缓冲区填满后,开发人员需要手动清空,然后才能继续使用。

8.1.1 启用 DBMS_OUTPUT

DBMS_OUTPUT 系统包将文本数据写入缓冲区,利用它把文本数据从程序保存到对应的缓存中,然后该缓存被其他的 PL/SQL 程序读取和操作。该系统包主要用于调试 PL/SQL 程序或在终端显示信息和报表。启用 DBMS_OUTPUT 系统包的命令如下。

```
DBMS_OUTPUT.ENABLE( buffer_size IN INTEGER DEFAULT 20000 );
```

参数说明:buffer_size 的默认值为 20000B,其值可设置的范围区间为[2000,1000000]。将 buffer_size 设置为 null,缓冲区大小将没有限制。

例如:

```
call dbms_output.enable();
call dbms_output.enable(NULL);
call dbms_output.enable(3000);
```

在没有启动 DBMS_OUTPUT 包情况下,编译器将忽略对其包下函数的调用。如果多次调用此函数,会以最后一次调用的 buffer_size 为准。当设置 buffer_size 为固定值时,一旦输入内容超过缓冲区的设置上限,程序将发生异常。开发人员往往将缓冲区大小设置为 NULL,确保程序的稳定运行。禁用 DBMS_OUTPUT 的语法格式如下。

```
DBMS_OUTPUT.DISABLE();
```

KSQL 启动和关闭 DBMS_OUTPUT 包的命令程序代码如下。

```
set serverout[put] on[;]
set serverout[put] off[;]
```

上述命令代码将向数据库服务器发出对应的 ENBALE(NULL)或 DISABLE 命令。

示例 8.1:启动与禁止 DBMS_OUTPUT。

程序代码如下。

```
create or replace procedure protest_A_1 is
BEGIN
  dbms_output.put_line('....test1');
  dbms_output.put_line('....test2');
END;
declare
  v_status integer := 0;
  v_data varchar2(100);
BEGIN
  call dbms_output.enable();
  protest_A_1();
  dbms_output.put_line('test3');
  dbms_output.get_line(v_data, v_status);
  dbms_output.put_line('v_data: ' ||v_data || 'v_status: ' ||v_status);
  dbms_output.put_line('test4');
  call dbms_output.disable();
END;
```

8.1.2 向缓冲区输入信息

可以通过调用 DBMS_OUTPUT.PUT、DBMS_OUTPUT.PUT_LINE 和 DBMS_OUTPUT.NEW_LINE 向缓冲区存入数据。语法格式如下。

将输入的数据追加到缓冲区末尾。

```
DBMS_OUTPUT.PUT(item IN VARCHAR2);
```

向缓冲区中添加新的一行数据。

```
DBMS_OUTPUT.PUT_LINE(item IN VARCHAR2);
```

向缓冲区添加一个换行符,产生新的一行。

```
DBMS_OUTPUT.NEW_LINE();
```

示例 8.2：使用 PUT、PUT_LINE、NEW_LINE 向缓冲区输入信息。
程序代码如下。

```
BEGIN
  call dbms_output.enable();
  dbms_output.put('test1');
  dbms_output.new_line();
  dbms_output.put('test2');
  dbms_output.new_line();
  dbms_output.new_line();
  dbms_output.put_line('test3');
  call dbms_output.disable();
END;
```

注意：若单行输入数据的大小超过限制，程序将报错。

8.1.3 从缓冲区读取信息

调用 DBMS_OUTPUT.PUT_LINE 等指令后，可以在终端中看到相应结果。接着可以调用 GET_LINE 和 GET_LINES 命令获取缓冲区中的信息。具体命令如下。

获取单行信息。GET_LINE 过程按照先入先出模式从缓存中提取一行信息，然后返回一个状态值，0 代表成功。程序代码如下。

```
DBMS_OUTPUT.GET_LINE(line OUT VARCHAR2, status OUT INTEGER);
```

获取多行信息。GET_LINES 过程在一个调用中从缓存中提取多行，把从缓存中读取的内容放在一个 PL/SQL 的字符串集合中。指明想要读取的行数，然后得到这些行，并传入对应的集合。

```
DBMS_OUTPUT.GET_LINES(lines OUT CHARARR, numlines IN OUT integer);
DBMS_OUTPUT.GET_LINES(lines OUT DBMSOUTPUT_LINESARRAY, numlines IN OUT integer)
```

参数说明如下。

（1）CHARARR 属于包内类型，定义为：TYPE CHARARR IS TABEL OF VARCHAR2(32767) INDEX BY INT。

（2）DBMSOUTPUT_LINESARRAY 属于包外类型，定义为：TYPE DBMSOUPUT_LINESARRAY IS VARRRAY(2147483647) OF VARCHAR2(32767)。

（3）Numlines 表示检索行数。当缓冲区行数大于等于检索行数时，返回检索行数；当缓冲区行数小于检索行数时，返回缓冲区行数。

以上两个重载方法的主要区别是，当传入检索行数小于零时，CHARARR 类型的方法不返回内容，而 DBMSOUTPUT_LINESARRAY 类型的方法运行异常。

示例 8.3：使用 GET_LINE、GET_LINES 从缓冲区读取信息。
程序代码如下。

```
DECLARE
  v_data dbms_output.CHARARR;
  numlines integer := 2;
BEGIN
  call dbms_output.enable();
  dbms_output.put_line('TEST 1');
  dbms_output.put_line('TEST 2');
  dbms_output.put_line('TEST 3');
  dbms_output.get_lines(v_data, numlines);
  dbms_output.put_line(v_data(1));
  dbms_output.put_line(v_data(2));
  call dbms_output.disable();
END;
```

8.2 文件读写

8.2.1 启动 UTL_FILE

UTL_FILE 包可以对操作系统中的文件进行读写操作。加载 UTL_FILE 插件的方式如下。

```
create extension utl_file;
```

相应地,卸载 UTL_FILE 插件的方式如下。

```
drop extension utl_file;
```

表 8.1 总结了 8 种常用方法来操作文件。

表 8.1 常用的操作文件的方法

方 法 名	作 用 描 述
FOPEN	打开一个文件,返回一个文件句柄
PUT_LINE	向文件中写入一行内容
PUT	将数据强制写入到文件中
FCLOSE	关闭目标文件
FCLOSE_ALL	关闭所有文件
GET_LINE	返回文件的一行
NEW_LINE	为文本增加行终止符
FFLUSH	清空目标文件

8.2.2 UTL_FILE 方法

1. FOPEN

FOPEN 用于打开一个文件,返回文件句柄,语法格式如下。

```
fopen(location text, filename text, open_mode text, max_linesize integer,
encoding name) return UTL_file.file_type
```

参数说明如下。

(1) Location：指定文件的位置。

(2) Filename：指定要打开的文件名。

(3) Open_mode：指定文件要打开的模式。一共有 R、W 和 A 三种模式，分别对应读取、写入和追加 3 种模式。

(4) Encoding：指定编码字符集。

2. PUT_LINE

PUT_LINE 的作用是向文件中写入一行内容，语法定义如下。

```
put_line(file file_type, buffer text  [, autoflush bool ]);
```

参数说明如下。

(1) File：其 file_type 类型就是打开文件后返回的句柄。

(2) Buffer：为输入的内容。

(3) Autoflush：表示是否自动刷新。

3. PUT

PUT 的作用和 PUT_LINE 类似，同样是向文件写入数据，不同的是 PUT 方法只是写入数据，不会在数据末尾加入换行符，这也是和 PUT_LINE 的主要区别。其语法定义如下。

```
put(file file_type, buffer text);
```

参数说明如下。

(1) File：其 file_type 类型就是打开文件后返回的句柄。

(2) Buffer：向文件中输入的内容。

4. FCLOSE

FCLOSE 的作用是关闭当前打开的文件，并释放相应的内存资源。其语法定义如下。

```
fclose(file file_type);
```

5. FCLOSE_ALL

FCLOSE 的作用是关闭所有打开的文件，并释放相应的内存资源。其语法定义如下。

```
utl_file.fclose_all();
```

6. GET_LINE

GET_LINE 的作用是获取文件中一行的数据。其语法定义如下。

```
get_line(file file_type, OUT buffer text [, len integer] );
```

参数说明如下。
(1) File：其 file_type 类型就是打开文件后返回的句柄。
(2) Buffer：获取后的数据。
(3) Len：指定想要输入的特定行。

7. NEW_LINE

NEW_LINE 的作用是向文本中增加一个换行符。其语法定义如下。

```
new_line(file file_type [, lines int] );
```

参数说明如下。
(1) File：目标文件。
(2) Lines：指定增加的行数。

8. FFLUSH

FFLUSH 的作用是将缓冲区的内容写入到文件。其语法定义如下。

```
fflush(file file_type);
```

其中，file 是将数据写入的目标文件。

示例 8.4：对文件信息进行手动复制。
程序代码如下。

```
CREATE or replace PROCEDURE proc1()
as
  DECLARE
    dirname         VARCHAR(100) DEFAULT '/tmp/';
    v_empfile_src   UTL_FILE.FILE_TYPE;
    v_empfile_tgt   UTL_FILE.FILE_TYPE;
    v_src_file      VARCHAR(20) DEFAULT 'source.txt';
    v_dest_file     VARCHAR(20) DEFAULT 'destination.txt';
    v_empline       VARCHAR(200);
BEGIN
    v_empfile_src = UTL_FILE.FOPEN(dirname,v_src_file,'r');
    v_empfile_tgt = UTL_FILE.FOPEN(dirname,v_dest_file,'w');
LOOP
    BEGIN
    UTL_FILE.GET_LINE(v_empfile_src,v_empline);
    raise notice 'v_empline: %', v_empline;
    UTL_FILE.PUT(v_empfile_tgt,v_empline);
    UTL_FILE.NEW_LINE(v_empfile_tgt);
    EXCEPTION
      WHEN OTHERS then
        EXIT;
    END;
```

```
    END LOOP;
    UTL_FILE.FCLOSE(v_empfile_tgt);
    UTL_FILE.FCLOSE(v_empfile_src);
END proc1;

CALL proc1();
```

程序运行结果如下。

```
cat /tmp/destination.txt
```

 ## 8.3 使用基于 Web 的数据(http)

如果想通过发送 http 请求包得到一些数据,可以有如下方法来实现。
(1) 通过 Web 浏览器获得。
(2) 使用类似 Python 的脚本语言。
(3) 通过类似 GUN 的 wget 的命令行工具。
(4) 使用 PL/SQL 的内置包 UTL_HTTP。

本节将探讨 http 包中一些最常用的数据类型及方法。可以通过 CREATE EXTENSION HTTP 导入 UTL_HTTP 和 http,以便将系统包加载到数据库实例中,允许数据库内使用包提供的方法检索网页信息。http 提供的方法相较于 UTL_HTTP,在程序代码编写上更为简洁明了,也更适合初学者上手。相应地,卸载命令为 DROP EXTENSION http。

8.3.1 UTL_HTTP 数据类型

当向服务器发送 http 请求时,需要先构建请求报文来发送请求数据,然后再构建相应的响应报文来接收服务器发出的响应。UTL_HTTP 系统包已经构建好如下的请求报文和响应报文的数据类型,定义如下。

```
CREATE TYPE UTL_HTTP.req AS(
url VARCHAR2(32767),
method VARCHAR2(64),
http_version VARCHAR2(64)
headers http_header[]);
```

参数说明如表 8.2 所示。

表 8.2　UTL_HTTP 的 req 参数说明

参　　数	描　　述
url	http 请求的 URL
method	对 URL 所标识的资源执行的方法,目前仅支持 GET

续表

参数	描述
http_version	http 协议版本
headers	请求头信息

```
CREATE TYPE UTL_HTTP.resp AS(
status_code INTEGER ,
reason_phrase VARCHAR2(256),
http_version VARCHAR2(64),
content TEXT);
```

参数说明如表 8.3 所示。

表 8.3 UTL_HTTP 的 resp 参数说明

参数	描述
status_code	Web 服务器返回的状态代码
reason_phrase	兼容性参数，无意义
http_version	http 协议版本
content	Web 服务器返回的内容

8.3.2 UTL_HTTP 方法

UTL_HTTP 系统包提供了如下方法发送 http 请求，向服务器获取信息。

(1) UTL_HTTP.BEGIN_REQUEST。

(2) UTL_HTTP.SET_HEADER。

(3) UTL_HTTP.GET_RESPONSE。

(4) UTL_HTTP.READ_LINE。

(5) UTL_HTTP.READ_TEXT。

(6) UTL_HTTP.END_RESPONSE。

(7) UTL_HTTP.ENDREQUEST。

以下介绍各个方法的使用场景、传参以及注意事项。

发起一个 http 请求的方法如下。

```
UTL_HTTP.BEGIN_REQUEST (
  url IN VARCHAR2,
  method IN VARCHAR2 DEFAULT 'GET',
  http_version IN VARCHAR2 DEFAULT NULL,
  request_context IN VARCHAR2 DEFAULT NULL,
  https_host IN VARCHAR2 DEFAULT NULL)
RETURN UTL_HTTP.req
```

参数说明如表 8.4 所示。

表 8.4　UTL_HTTP 的 BEGIN_REQUEST 参数说明

参　　数	描　　述
url	http 请求的 URL
method	对 URL 所标识的资源执行的方法,目前仅支持 GET
http_version	http 协议版本
request_context	请求体
https_host	兼容性参数,无意义

此方法返回值对象是 UTL_HTTP.req 对象。

http 头字段(HTTP header fields)是指在超文本传输协议(http)的请求和响应消息中的消息头部分,定义了一个超文本传输协议事务中的操作参数,http 头部字段可以根据需要定义。因此,在 Web 服务器和浏览器上发现非标准的头字段,可以通过配合请求头和响应头实现协商缓存、会话状态等功能。在 UTL_HTTP 系统包中,对响应头的设置方法如下。

```
UTL_HTTP.SET_HEADER (
  r IN OUT UTL_HTTP.req,
  name IN VARCHAR2,
  value IN VARCHAR2
```

参数说明如表 8.5 所示。

表 8.5　UTL_HTTP 的 SET_HEADER 参数说明

参　　数	描　　述
r	http 的请求
name	http 请求头的名字
value	http 请求头的值

一次完整的 HTTP 请求过程包括发送请求报文和响应报文两个步骤,上述操作已经包含了发送操作,接收响应报文的方法如下。

```
UTL_HTTP.GET_RESPONSE (
  r IN UTL_HTTP.req,
  return_info_response IN BOOLEAN DEFAULT FALSE)
  RETURN UTL_HTTP.resp
```

参数说明如表 8.6 所示。

表 8.6　UTL_HTTP 的 GET_RESPONSE 参数说明

参　　数	描　　述
R	http 的请求
return_info_response	兼容性参数,无意义

此方法将返回 UTL_HTTP.resp 类型的响应报文，具体的响应参数以及响应体等信息都可以通过此对象获得。

UTL_HTTP.READ_LINE 方法以文本形式读取 http 响应体，直到到达行尾，并在调用者提供的缓冲区返回输出。如果到达响应正文的末尾，将引发 NO_DATA_FOUND 异常。方法如下。

```
UTL_HTTP.READ_LINE(
  r IN OUT UTL_HTTP.resp,
  data OUT VARCHAR2,
  remove_crlf IN BOOLEAN DEFAULT FALSE)
```

参数说明如表 8.7 所示。

表 8.7　UTL_HTTP 的 READ_LINE 参数说明

参数	描述
r	http 的响应
data	http 的响应正文
remove_crlf	兼容性参数，无意义

UTL_HTTP.READ_TEXT 方法以文本形式读取 http 响应体，并在调用者提供的缓冲区返回输出。如果到达响应正文的末尾，将引发 NO_DATA_FOUND 异常。在传入参数中，如果 len 值传入为 null，该方法将读取尽可能多的输入来填充数据。方法如下。

```
UTL_HTTP.READ_TEXT(
  r IN OUT UTL_HTTP.resp,
  data OUT VARCHAR2,
  len IN INTEGER DEFAULT NULL)
```

参数说明如表 8.8 所示。

表 8.8　UTL_HTTP 的 READ_TEXT 参数说明

参数	描述
r	http 的响应
data	http 的响应正文
len	要读取数据的最大字符数

一次 http 请求需要以消耗系统资源为代价建立起连接。为减少系统的不必要开销，使用完后应关闭连接。关闭的方法如下。

```
--此过程将结束 HTTP 响应
UTL_HTTP.END_RESPONSE ( r IN OUT UTL_HTTP.resp)
--此过程将结束 HTTP 请求
UTL_HTTP.END_REQUEST ( r IN OUT UTL_HTTP.req)
```

示例 8.5：使用 UTL_HTTP 方法访问金仓数据库官网，并获取响应。

程序代码如下。

```
DECLARE
  req UTL_HTTP.REQ;
  resp UTL_HTTP.RESP;
    buf varchar2(32767);
BEGIN
  req := UTL_HTTP.BEGIN_REQUEST('https://www.kingbase.com.cn/','GET','HTTP/1.1');
  UTL_HTTP.SET_HEADER(req, 'User-Agent','Mozilla/5.0');
  resp := UTL_HTTP.GET_RESPONSE(req);
  RAISE NOTICE 'response status code:%', resp.STATUS_CODE;
  RAISE NOTICE 'Contents of the response are as follow:';
  BEGIN
    LOOP
      UTL_HTTP.READ_LINE(resp, buf);
      RAISE NOTICE 'content:%', buf;
    END LOOP;
  EXCEPTION
    WHEN NO_DATA_FOUND
    THEN
      NULL;
  END;
  UTL_HTTP.END_RESPONSE(resp);
  UTL_HTTP.END_REQUEST(req);
END;
```

程序运行结果如下。

```
response status code:200
Contents of the response are as follow:
content:<!DOCTYPE html>
content:<html lang="en">
content:<head>
content:    <meta charset="UTF-8">
content:    <meta http-equiv="X-UA-Compatible" content="IE=edge">
content:    <meta name="viewport" content="width=device-width, initial-scale=1.0">
content:    <title>人大金仓-成为世界卓越的数据库产品与服务提供商</title>
content:    <meta name="keywords" content="云数据库,MPP 数据库,自主可控数据库,国产数据库,国产基础软件,国产化替代,Oracle 兼容,Oracle 替代,电力系统数据库">
......
```

8.3.3　http 数据类型

1．http_request

```
CREATE TYPE http_request AS (
  method http_method,
```

```
    uri VARCHAR,
    headers http_header[],
    content_type VARCHAR,
    content VARCHAR
);
```

参数说明如表 8.9。

表 8.9　http_request 的参数说明

参　　数	描　　述
method	对 URL 所标识的资源执行的方法
uri	http 请求的 URL
headers	请求头信息
content_type	请求类型
content	请求内容

2. http_response

```
CREATE TYPE http_response AS (
    status INTEGER,
    content_type VARCHAR,
    headers http_header[],
    content VARCHAR
);
```

参数说明如表 8.10。

表 8.10　http_response 的参数说明

参　　数	描　　述
status	Web 服务器返回的状态代码
content_type	Web 服务器返回内容的类型
headers	Web 服务器返回的 HTTP 头信息
content	Web 服务器返回的内容

3. http_header

```
CREATE TYPE http_header AS (
    field VARCHAR,
    value VARCHAR
);
```

参数说明如表 8.11 所示。

表 8.11 http_header 的参数说明

参　　数	描　　述
field	http 请求头的名字
value	http 请求头的值

4. http_method

```
CREATE DOMAIN http_method AS text
CHECK (
    VALUE ILIKE 'get' OR
    VALUE ILIKE 'post' OR
    VALUE ILIKE 'put' OR
    VALUE ILIKE 'delete' OR
    VALUE ILIKE 'patch' OR
    VALUE ILIKE 'head'
);
```

8.3.4　http 方法

http 提供了 GET、POST、PUT、DELETE 等常用方法,用于向服务器发起请求,以便获取所需的资源。

1. http 方法

语法格式如下。

```
http(request http_request) returns http_response
```

示例 8.6：使用 http 方法发送请求,并获取响应。

程序代码如下。

```
SELECT status,
    content::json->'args' as args,
    content::json->>'data' as data,
    content::json->'url' as url,
    content::json->'method' as method
    from http(('GET', 'https://httpbin.org/anything', NULL, 'application/json
', '{"search": "toto"}'));
```

程序运行结果如下。

```
status | args | data            | url                           | method |
-----+-----+-----------------+-------------------------------+------+--
 200   | {}  | {"search": "toto"} | "https://httpbin.org/anything"| "GET"|
```

2. http_get 方法

语法格式如下。

```
http_get(uri VARCHAR) returns http_response
```

示例 8.7：使用 http_get 方法发送请求，并获取响应。
程序代码如下。

```
SELECT content FROM http_get('http://httpbin.org/ip');
```

程序运行结果如下。

```
content                     |
----------------------------+
{"origin":"24.69.186.43"}   |
```

3. http_post 方法
语法格式如下。

```
http_post(uri VARCHAR, content VARCHAR, content_type VARCHAR)
returns http_response
```

示例 8.8：使用 http_post 方法发送请求，并获取响应。
程序代码如下。

```
SELECT status,
    content::json->'data' AS data,
    content::json->'args' AS args,
    content::json->'url' AS url,
    content::json->'method' AS method
FROM
http_post('https://httpbin.org/anything?foo=bar','payload','text/plain');
```

程序运行结果如下。

```
status | data      | args            | url                                      | method |
-------+-----------+-----------------+------------------------------------------+--------+--
200    | "payload" | { "foo":"bar" } | "https://httpbin.org/anything?foo=bar"   | "POST" |
```

4. http_put 方法
语法格式如下。

```
http_put(uri VARCHAR, content VARCHAR, content_type VARCHAR)
returns http_response
```

示例 8.9：使用 http_put 方法发送请求，并获取响应。
程序代码如下。

```
SELECT status, content_type, content::json->>'data' AS data
  FROM http_put('http://httpbin.org/put', 'some text', 'text/plain');
```

程序运行结果如下。

```
status  | content_type     | data      |
--------+------------------+-----------+
200     | application/json | some text |
```

5．http_delete 方法

语法格式如下。

```
http_delete(uri VARCHAR) returns http_response
```

示例 8.10：使用 http_delete 方法发送请求，并获取响应。

程序代码如下。

```
SELECT status,
content::json->'args' AS args,
content::json->'url' AS url,
content::json->'method' AS method
FROM http_delete('https://httpbin.org/anything?foo=bar');
```

程序运行结果如下。

```
status  | args            | url                                   | method |
--------+-----------------+---------------------------------------+--------+
200     | {"foo": "bar"}  | "https://httpbin.org/anything?foo=bar"| "DELETE"|
```

示例 8.11：创建使用 http、http_get 和 http_post 方法向指定 URI 发送请求并获取响应的存储过程，然后调用存储过程。

程序代码如下。

```
CREATE OR REPLACE PROCEDURE getHttpResponse1 (IN uri varchar2(100))
AS
DECLARE
    data1 varchar2(2000);
BEGIN
    if uri='https://httpbin.org/anything' then
        data1:=http(('GET', 'https://httpbin.org/anything', NULL, 'application/json', '{"search": "toto"}'));
    elsif uri='http://httpbin.org/ip' then
        data1:=http_get('http://httpbin.org/ip');
    elsif uri='https://httpbin.org/anything?foo=bar' then
        data1:=http_post('https://httpbin.org/anything?foo=bar','payload','text/plain');
    end if;
```

```
        raise notice '%', data1;
END;

CALL getHttpResponse1('http://httpbin.org/ip');
```

程序运行结果如下。

```
(200,application/json,"{""(Date, \""Sat,? 08? Jul? 2023? 03:20:30? GMT\"")"",""
(Content-Type, application/json)"","" (Content- Length, 31)"","" (Connection,
close)"","" (Server, gunicorn/19.9.0)"","" (Access - Control - Allow - Origin,
* )"",""(Access-Control-Allow-Credentials,true)""}","{
??""origin"":?""175.162.8.95""
}")
```

第 9 章

PL/SQL 子程序

PL/SQL 子程序是可以重复调用的命名 PL/SQL 块。子程序可以是过程，也可以是函数。通常，使用过程来执行操作，使用函数来计算并返回结果值。本章包括以下主要内容。

(1) 子程序概述。
(2) 独立子程序。
(3) 嵌套子程序。
(4) 子程序重载。
(5) 表函数。

 ## 9.1　子程序概述

9.1.1　子程序的分类

PL/SQL 子程序是命名的 PL/SQL 块，可以反复调用，体现了 PL/SQL 程序的模块化，包括函数和过程两种。其中，函数用于返回特定的数据，过程用于执行特定的操作，不需要返回值。调用时，过程可以作为一个独立的表达式被调用，而函数只能作为表达式的一个组成部分被调用。

可以在 PL/SQL 块（该块也可以是一个子程序）或 PL/SQL 包中定义子程序。根据子程序的存在形式不同，PL/SQL 中的子程序可分为独立子程序、嵌套子程序、包子程序。

（1）独立子程序：在数据库模式下创建的子程序称为独立子程序。独立子程序存储在数据库服务器中，可以通过 CREATE FUNCTION 或 CREATE PROCEDURE 语句创建。

（2）嵌套子程序：在一个 PL/SQL 块中创建的子程序称为嵌套子程序。可以同时进行子程序的声明与定义，也可以先声明子程序，然后定义该子程序。

（3）包子程序：在 PL/SQL 包中创建的子程序称为包子程序。可以在包规范中声明子程序，在包体中实现子程序。包子程序随包一起存储在数据库服务器中。包子程序详见第 12 章。

9.1.2　子程序的优点

PL/SQL 子程序的使用将有助于提高程序开发和维护的可靠性，以及代码的可重用性，具体内容如下。

（1）模块化。

子程序可以将大的代码块程序分解为可管理的、若干较小、定义良好的模块，这些模块又可以被其他模块调用。

（2）简化应用程序设计。

设计应用程序时，可以先考虑主体，然后再考虑子程序的具体实现细节（通过在子程序中使用 NULL 语句），直到测试了主程序，然后逐步细化它们。

（3）可维护性。

可以更改子程序的实现细节，而不需要更改其调用程序。

（4）可包装性。

子程序可以分组到包中，易于维护与管理。

（5）可重用性。

采用"即插即用"的原则，在许多不同的环境中，将不同的包子程序或独立子程序进行组合，可以实现想要的功能。

（6）更好的性能。

每个子程序都以可执行的形式编译和存储，在用户之间共享，仅在需要的时候才使用，并且可以重复调用。这种机制减少了数据在数据库和应用服务器之间的传输，并缩短了响应时间，同时存储的子程序被缓存，也降低了内存需求和调用开销。

9.2 独立子程序

9.2.1 子程序结构

PL/SQL 子程序由 4 部分组成，如图 9.1 所示。

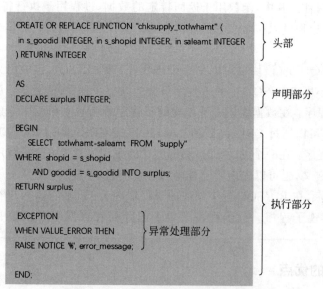

图 9.1 PL/SQL 子程序的 4 部分组成

具体内容如下。

(1) 子程序头部：指定过程或函数的名称、参数名称与类型以及函数的返回值类型。

(2) 声明部分：声明和定义局部变量、游标、常量、异常以及嵌套子程序等，除游标外，这些元素的生命周期仅存在于子程序内部。

(3) 执行部分：由一个或多个语句组成，包括赋值语句、控制语句、DML 语句等。

(4) 异常处理部分：处理子程序运行时产生的异常。

9.2.2 创建函数

调用函数前，必须声明和定义它。可以先声明它，然后在同一个块、子程序或包中定义它，或者同时声明和定义它。

函数是有返回值的子程序，值的数据类型是函数的数据类型。函数调用是一个表达式，其数据类型是函数的数据类型。

1. 创建函数

函数的通用创建形式如下。

```
create_function ::=
CREATE [ OR REPLACE ] [ EDITIONABLE | NONEDITIONABLE ]
FUNCTION plsql_function_source
plsql_function_source ::=
[ schema.] function_name
[ ( parameter_declaration [, parameter_declaration]... ) ] RETURN datatype
[ { invoker_rights_clause
| pipelined_clause
}...
]
{ IS | AS }
{ [ declare_section ] body }
```

语义描述如下。

(1) CREATE[OR REPLACE]。

如果存在，则重新创建该函数，并重新编译它。

重新定义函数之前被授予权限的用户仍然可以访问该函数，而无须重新授予权限。

(2) [EDITIONABLE | NONEDITIONABLE]：指明函数是否允许被编辑。默认为 EDITIONABLE。

(3) schema：包含函数的模式名称。默认值为当前模式。

(4) function_name：要创建的函数名称。

(5) RETURN datatype。

对于 datatype，指定函数返回值的数据类型。返回值可以是 PL/SQL 支持的任何数据类型。数据类型不能指定长度、精度或小数位数。数据库从调用函数的环境中获取返回值的长度、精度或小数位数。不能限制此数据类型，如使用 NOT NULL。

(6) body 是函数所需的可执行部分，以及函数的异常处理部分(可选)。

(7) declare_section 是函数的可选声明部分。声明是函数本地的，可以在 body 中引用,并且在函数完成执行时不再存在。

2. 修改函数

ALTER FUNCTION 可以更改一个函数的定义,请参考 KingbaseES PL/SQL 过程语言参考手册。

3. 删除函数

DROP FUNCTION 可以删除一个函数的定义,请参考 KingbaseES PL/SQL 过程语言参考手册。

4. 用户自定义函数举例

示例 9.1:函数的创建。

功能描述:创建一个函数,根据参数指定的货物编码、商店编码、卖出货物数量,计算库存减去指定的货物数量后的剩余库存数量。

程序代码如下。

```
--创建名为 chksupply_totlwhamt 的函数,如果存在则修改
--依次定义了货物编码、商店编码、卖出货物数量 3 个参数,数据类型为整数
CREATE OR REPLACE FUNCTION "chksupply_totlwhamt" (
in s_goodid INTEGER, in s_shopid INTEGER, in saleamt INTEGER
) RETURNs INTEGER as              --定义了函数返回类型为整数
DECLARE surplus INTEGER;          --声明了名为 surplus 的整数变量
BEGIN
  SELECT  totlwhamt-saleamt INTO surplus  FROM  "supply"
  WHERE   shopid = s_shopid AND goodid = s_goodid;
  --查询指定商店的指定货物的库存数量,并减去传入的卖出货物数量
  RETURN surplus;--返回剩余库存数量
END;
```

5. 利用图形化界面创建函数

可以在 KStudio 中利用图形化界面来辅助创建函数。如图 9.2 所示,在基本属性中输入参数名称、返回值类型以及模式,在参数列表中输入需要的参数,包括数据类型,在 SQL 内容中输入具体的执行部分和异常处理部分。

9.2.3 创建存储过程

存储过程是存储在数据库中的过程。创建一个新过程或者替换一个已有的定义,需要使用 CREATE OR REPLACE PROCEDURE 语句。为了能够定义过程,用户需具有所使用语言上的 USAGE 特权。

如果这个命令中包括一个模式名称,则该过程将被创建在此模式中。否则,过程将被创建在当前的模式中。新过程的名称不能和同一模式中具有相同输入参数类型的任何现有过程或函数同名。不过,具有不同参数类型的过程和函数可以同名(称为重载)。

当 CREATE OR REPLACE PROCEDURE 被用来替换一个现有的过程时,该过程的拥有关系和权限保持不变。所有其他的过程属性会被赋予这个命令中指定的或隐含的值。必须拥有(包括成为拥有它的角色)该过程才能替换它。

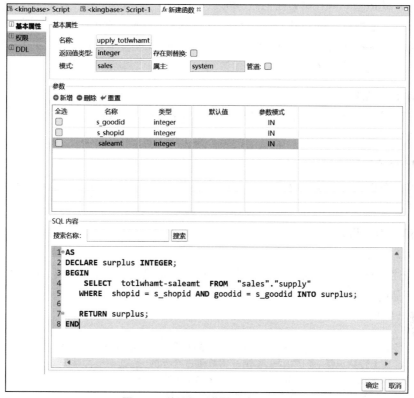

图 9.2 利用图形化界面创建函数

1. 创建存储过程

存储过程的通用创建形式如下。

```
create_procedure ::=
CREATE [ OR REPLACE ] [ EDITIONABLE | NONEDITIONABLE ]
PROCEDURE plsql_procedure_source

plsql_procedure_source ::=
[ schema. ] procedure_name
[ ( parameter_declaration [, parameter_declaration ]... ) ]
[ invoker_rights_clause ]
{ IS | AS }
{ [ declare_section ] body } ;
```

语义描述如下。

(1) CREATE[OR REPLACE]

如果存在,则重新创建该过程,并重新编译它。

重新定义过程之前被授予权限的用户仍然可以访问该过程,而无须重新授予权限。

(2)[EDITIONABLE｜NONEDITIONABLE]:指明存储过程是否允许被编辑。默认为 EDITIONABLE。

(3) schema:过程的模式的名称。默认值为当前模式。

(4) procedure_name：要创建的过程的名称。

(5) body：过程所需的可执行部分，以及（可选）过程的异常处理部分。

(6) declare_section：过程的可选声明部分。声明是过程本地的，可以在主体中引用，并且在过程完成执行时不再存在。

2. 修改存储过程

ALTER PROCEDURE 可以更改一个存储过程的定义，请参考 KingbaseES PL/SQL 过程语言参考手册。

3. 删除存储过程

DROP PROCEDURE 可以删除一个存储过程的定义，请参考 KingbaseES PL/SQL 过程语言参考手册。

4. 存储过程应用举例

示例 9.2：存储过程的创建。

功能描述：创建一个存储过程，根据参数指定的货物编码、商店编码、卖出货物数量，更新该指定商品的库存数量。

程序代码如下。

```
--创建名为 updatesupply_totlwhamt 的存储过程,如果存在则修改
--依次定义了货物编码、商店编码、卖出货物数量 3 个参数,数据类型为整数
CREATE OR REPLACE PROCEDURE "updatesupply_totlwhamt" (
    IN s_goodid INTEGER, IN s_shopid INTEGER, IN s_saleamt INTEGER
) AS
DECLARE surplus INTEGER; --声明了名为 surplus 的整数变量
BEGIN
    --调用上一节创建的 chksupply_totlwhamt 函数,并将返回值赋值给 surplus 变量
    surplus:="chksupply_totlwhamt"(s_goodid,s_shopid,saleamt);
    IF surplus>0--判断 surplus 大于 0 则执行更新操作
    THEN
        UPDATE "supply" SET "totlwhamt" = "totlwhamt" -s_saleamt
        WHERE shopid = s_shopid AND goodid = s_goodid;
    END IF;--结束 IF 判断
END;
```

5. 利用图形化界面创建存储过程

可以在 KStudio 中利用图形化界面来辅助创建存储过程。如图 9.3 所示，在基本属性中输入存储过程名称以及模式，在参数列表中输入需要的参数，包括数据类型，在 SQL 内容中输入具体的执行部分和异常处理部分。

9.2.4 支持的参数

KingbaseES 存储模块可以接受 KingbaseES 数据库系统所支持的任何数据类型，作为输入输出参数和声明的变量类型。KingbaseES 存储过程和函数提供了 IN、OUT、IN OUT 3 种参数模式，分别对应输入、输出、输入输出 3 种语义。IN 参数将值传给被调用的子程序；OUT/INOUT 参数调用时需传入变量，调用后实参变量的值将被修改。不声明参数模

第 9 章 PL/SQL 子程序

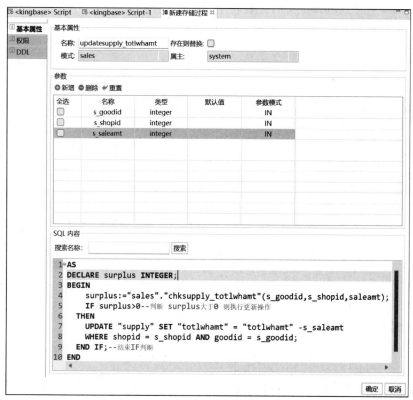

图 9.3 利用图形化界面创建存储过程

式时,默认为 IN 类型。KingbaseES 支持模式的参数与主流的数据库和 SQL 的标准规范对比如表 9.1 所示。

表 9.1 各数据库支持的输入输出参数对比

	函数带有 OUT 参数	函数的返回值	存储过程带有 OUT 参数	存储过程的返回值
SQL 标准	NO	YES	YES	NO
SQLServer	NO	YES	YES	NO
Oracle	YES	YES	YES	NO
KingBaseES	YES	YES	YES	NO

其中,YES 表示允许或有,NO 表示禁止或没有。

如果有 OUT 或 INOUT 参数,运行所得是一个结果集,结果集由一条或多条 RECORD 组成。在每条 RECORD 中,字段的顺序按 OUT 或 INOUT 参数声明的顺序。如果函数或存储过程的参数有 OUT 或 INOUT 参数,函数或存储过程的语句中不能带有返回结果语句,如 RETURN 语句、RETURN NEXT 语句等。

参数的指定允许以位置参数或命名参数两种格式进行,如果同时使用位置参数和命名参数,命名参数必须出现在位置参数之后。命名参数主要用于调用存储过程具有默认值的情况,OUT/IN OUT 参数不能设置默认值。利用命名参数的方式调用存储过程时,参数的出现顺序不需要与存储过程定义的参数顺序相同。

以代码示例 9.1 为例(函数见示例 9.1)执行该函数,程序代码如下。

```
DECLARE
  RETURNVar INTEGER;
  s_goodid BIGINT;
  s_shopid BIGINT;
  saleamt INTEGER;                    --声明参数
BEGIN
  s_goodid := 12560557;
  s_shopid := 1000004142;
  saleamt := 2;                       --参数赋值
  --调用示例 9.1 创建的函数
  RETURNVar:= "chksupply_totlwhamt"(s_goodid, s_shopid, saleamt);
  RAISE NOTICE '%', RETURNVar;
END;
  NOTICE: 998
```

示例 9.3:在上面函数的基础上增加 OUT 参数。

程序代码如下。

```
CREATE OR REPLACE FUNCTION "chksupply_totlwhamt2" (
  IN s_goodid BIGINT, IN s_shopid BIGINT, IN saleamt INTEGER, OUT result INTEGER-
-在原先三个 IN 参数的基础上增加了第四个 OUT 参数
) RETURNS INTEGER AS
DECLARE surplus INTEGER;
BEGIN
  SELECT totlwhamt-saleamt FROM "supply"
   WHERE shopid = s_shopid AND goodid = s_goodid INTO surplus;
result:=surplus;--将结果 surplus 赋值给了前面定义的 OUT 参数
RETURN result;
END;
```

执行该函数,程序代码如下。

```
DECLARE
  RETURNVar INTEGER;
  s_goodid BIGINT;
  s_shopid BIGINT;
  saleamt INTEGER;
  result INTEGER;
BEGIN
  s_goodid := 12560557;
  s_shopid := 1000004142;
  saleamt := 2;
  RETURNVar:= "chksupply_totlwhamt2"(s_goodid, s_shopid, saleamt, result);
  RAISE NOTICE '%, %', RETURNVar, result;
END;
```

程序运行结果如下。

```
NOTICE: 998,998
```

示例 9.4：继续将该函数改为带 INOUT 参数的函数。

程序代码如下。

```
CREATE OR REPLACE FUNCTION " chksupply_totlwhamt3" (
  IN s_goodid BIGINT, IN s_shopid BIGINT, IN OUT saleamt INTEGER, result INTEGER-
--将原先第三个 IN 参数改为 INOUT 参数
) RETURNS INTEGER AS
DECLARE surplus INTEGER;
BEGIN
  SELECT totlwhamt-saleamt FROM "supply"
  WHERE shopid = s_shopid AND goodid = s_goodid INTO surplus;
saleamt :=surplus;--将结果 surplus 赋值给了前面定义的 INOUT 参数
RETURN saleamt ;
END;
```

执行该函数，程序代码如下。

```
DECLARE
   RETURNVar INTEGER;
   s_goodid BIGINT;
   s_shopid BIGINT;
   saleamt INTEGER;
BEGIN
   s_goodid :=12560557;
   s_shopid :=1000004142;
   saleamt :=2;
   RETURNVar:= "chksupply_totlwhamt3"(s_goodid,s_shopid,saleamt);
   RAISE NOTICE '%, %', RETURNVar, saleamt;
END;
```

程序运行结果如下。

```
NOTICE: 998,998
```

9.2.5 调用与使用

创建独立子程序后，它以编译的形式存储于数据库服务器端，供应用程序调用。如果不调用，则独立子程序不被执行。通过独立子程序名称调用独立子程序时，实参的数量、顺序、类型要与形参的数量、顺序、类型相匹配。

子程序调用具有以下形式。

```
subprogram_name [ ( [ parameter [, parameter]... ] ) ]
```

如果子程序没有参数，或为每个参数指定一个默认值，则可以省略该参数列表或指定一个空的参数列表。一个过程调用一个 PL/SQL 语句,一个函数调用一个表达式。

存储过程可以使用 CALL 调用,用户自定义函数可以通过 SELECT 语句调用。对于存储过程,由于不能和其他任何常量、函数、存储过程等一并构成表达式使用,因此只能单独作为一个表达式出现在 SELECT 语句中。而对于用户自定义函数,如果没有 OUT 或 INOUT 参数,可以和其他常量、变量、对象名如字段名等组合成表达式使用。带有 OUT 或 INOUT 参数的函数不能参与表达式的计算。如果一个函数返回值是结果集,可以使用如下格式获取返回值。

```
SELECT * FROM funname(...);--funname 为自定义函数名称
```

函数可以出现在表达式中,如

```
CREATE FUNCTION increment(i INTEGER) RETURNS INTEGER AS BEGIN RETURN i + 1; END;
```

SELECT 可以调用并用于表达式计算,如

```
SELECT increment(1) + increment(10) AS result;
```

程序运行结果如下。

```
RESULT
--------
13
```

有 OUT 参数时,不正确的使用方式如下。

```
CREATE FUNCTION increment2(i OUT INTEGER) RETURNS INTEGER AS
  BEGIN
    i := 1;
RETURN 4;
END;
```

SELECT 调用的方式如下。

```
SELECT increment2(1);
ERROR: proc increment2 has OUT/IN OUT parameter, unresolved action.
```

通过 SELECT 获取返回结果或者结果集如下。

```
CREATE OR REPLACE FUNCTION mypower(i INTEGER) RETURNS SETOF INTEGER AS
BEGIN
  FOR j IN 1..i LOOP
  RETURN NEXT j * j;
    END LOOP;
END;
```

SELECT 调用的方式如下。

```
SELECT * FROM mypower(3);
```

程序运行结果如下。

```
MYPOWER
---------
    1
    4
    9
```

存储过程通过 CALL 执行。如果过程有任何输出参数,则会返回一个结果行,返回这些参数的值,程序代码如下。

```
--创建一个存储过程
CREATE OR REPLACE PROCEDURE   getfirstgoodname()
AS
BEGIN
  SELECT  TOP 1 goodname
  FROM   "goods";
END;
```

CALL 调用的方式如下。

```
CALL " getfirstgoodname "();
```

程序运行结果如下。

```
goodname
---------
```

9.2.6 支持的返回值类型

KingbaseES 存储过程和函数可以返回 KingbaseES 数据库系统支持的任何数据类型作为返回值;另外,还可以返回如下一些类型作为返回值。

(1) Refcursor:引用游标类型。

(2) SETOF datatype:集合类型,通常需要配合 RETURN NEXT … 语句使用,datatype 可以是 KingbaseES 数据库系统支持的任何数据类型,甚至是一个已存在的表的名称。但 SETOF 不能和 OUT/IN OUT 参数一起使用。

1. 返回值为 SETOF INTEGER 类型函数

示例 9.5:将示例 9.1 改造为返回值为 SETOF INTEGER 的类型函数。
程序代码如下。

```
CREATE OR REPLACE FUNCTION "chksupply_totlwhamt_setof" (
IN s_goodid BIGINT, IN s_shopid BIGINT, IN saleamt INTEGER
```

```
                                            --定义了3个IN参数变量
) RETURNS SETOF INTEGER AS
DECLARE _totlwhamt INTEGER;
BEGIN
  SELECT
  totlwhamt
  FROM
  "supply"
  WHERE
  shopid = s_shopid
AND goodid = s_goodid INTO _totlwhamt;      --获取库存量

RETURN NEXT _totlwhamt;                     --返回库存数量
RETURN NEXT saleamt;                        --返回卖出数量
RETURN NEXT _totlwhamt-saleamt;             --返回剩余库存数量
END;
```

执行该函数,程序代码如下。

```
DECLARE

  s_goodid BIGINT;
  s_shopid BIGINT;
  saleamt INTEGER;
BEGIN
  s_goodid :=12560557;
  s_shopid :=1000004142;
  saleamt :=2;
SELECT * FROM "chksupply_totlwhamt_setof"(s_goodid,s_shopid,saleamt);

END;
```

程序运行结果如下。

```
chksupply_totlwhamt_setof
---------
2
1000
998
```

2. 返回值为 SETOF TABLE 函数

示例 9.6:查询某指定商店商品的列表函数。

程序代码如下。

```
CREATE OR REPLACE FUNCTION getGoodlistByShop (IN s_shopid BIGINT) RETURN SETOF "
supply" AS --"supply"是库存表表名,请参见示例数据库

DECLARE
CURSOR cc for SELECT * FROM "supply" WHERE
```

```
      shopid = s_shopid;
      rs RECORD;
BEGIN
    OPEN cc;
   LOOP
     FETCH cc INTO rs;
     EXIT WHEN cc%NOTFOUND;
     RETURN NEXT rs;
   END LOOP;
   CLOSE cc;
END;
```

执行该函数,程序代码如下。

```
DECLARE
   s_shopid BIGINT;
BEGIN
   s_shopid:=1728689939;
     SELECT * FROM"getgoodlistbyshop"(s_shopid);
END;
```

程序运行结果如下。

```
(104142,"12560557 ",1000,$116.80,0.99,t,https://item.jd.com/12560557.html,)
```

3. REFCURSOR 作为函数参数和返回值

示例 9.7:REFCURSOR 作为函数参数和返回值。

程序代码如下。

```
--创建一个返回值类型为REFCURSOR的用户自定义函数
CREATE OR REPLACE FUNCTION getGoodlistFun(IN s_shopid BIGINT)
RETURN REFCURSOR AS
DECLARE
    CURSOR c1 FOR SELECT * FROM "goods";
BEGIN
    OPEN c1;
    RETURN c1;
END;
--调用上面创建的函数,创建一个使用返回类型为REFCURSOR的存储过程
CREATE OR REPLACE PROCEDURE getgoodlistpro()
AS
DECLARE
    c2 REFCURSOR;
    rs RECORD;
BEGIN
    c2 := "getGoodlistFun"();
    FETCH c2 INTO rs;
    RAISE NOTICE 'name:%', rs.goodname;
END;
```

程序运行结果如下。

```
NOTICE: name:知识图谱:方法、实践与应用(博文视点出品)
```

4. 返回结果集

服务器支持存储过程或匿名块返回结果集给客户端。支持返回结果集的语句如下：SELECT 语句(而不是 SELECT..INTO)。

示例 9.8

程序代码如下。

```
CREATE OR REPLACE PROCEDURE getGoodlistbyname()
AS
DECLARE
goodname2 RECORD;
BEGIN
    SELECT  *  FROM "goods"
    WHERE  goodname LIKE '%小松鼠%'  ;
END;
```

执行该存储过程,程序代码如下。

```
CALL  getGoodlistbyname();
```

程序运行结果如下。

```
 goodname        | goodid        | mfrs              | brand  | ...
-----------------+---------------+-------------------+--------+---
 小松鼠蜀香牛肉   | 571511639661  | 罗罗牛肉食品有限公司 | 小松鼠 | ...
 小松鼠菠萝干     | 551203237670  | 小松鼠有限公司      | 小松鼠 | ...
 小松鼠手撕鱿鱼片 | 551203237670  | 富美食品有限公司    | 小松鼠 | ...
 小松鼠碧根果     | 562395925542  | 小松鼠有限公司      | 小松鼠 | ...
 小松鼠手剥巴旦木 | 562465178375  | 小松鼠有限公司      | 小松鼠 | ...
 小松鼠芒果干     | 45735764241   | 牡丹食品有限公司    | 小松鼠 | ...
 小松鼠夏威夷果   | 560823257520  | 小松鼠有限公司      | 小松鼠 | ...
 小松鼠_酸辣粉    | 606326339839  | 四川阿满食品有限公司 | 小松鼠 | ...
```

使用 OUT 或 INOUT 参数和返回值的注意事项如下。

(1) 函数和存储过程都可以有 OUT 参数或 INOUT 参数。

(2) 函数有 OUT 参数或 INOUT 参数时,返回值不能是集合,即不能为"RETURN SETOF"。

(3) 调用带有 OUT 参数的存储过程或函数时,OUT/IN OUT 参数只能接收变量。

9.3 嵌套子程序

9.3.1 概述

在 PL/SQL 块内创建的子程序称为嵌套子程序。可以同时声明和定义它,也可以先声明它,然后在同一块中定义它。只有嵌套在独立子程序或包子程序中的嵌套子程序才会存储在数据库中。

嵌套子程序对外部程序不可见。

9.3.2 声明和定义

声明了嵌套子程序,但是没有进行定义,称为前置声明,可以稍后在同一个 PL/SQL 块中定义它。其中,函数声明的语法格式如下。

```
FUNCTION name ( [ [ argmode ] [ argname ] argtype [ { DEFAULT | = } default_expr ]
[,...] ] )
[ RETURNS rettype];
```

函数定义的语法格式如下。

```
function_heading IS | AS
...;
```

其中,IS 或 AS 后面是一个 PL/SQL 块。上面语法中的参数信息参见独立子程序中的 CREATE FUNCTION。

存储过程声明的语法格式如下。

```
PROCEDURE name ( [ [ argmode ] [ argname ] argtype [ { DEFAULT | = }
default_expr ] [,...] ] );
```

存储过程定义的语法格式如下。

```
procedure_declaration IS | AS
...;
```

其中,IS 或 AS 后面是一个 PL/SQL 块。语法中的参数信息参见独立子程序中的 CREATE PROCEDURE。

示例 9.9:嵌套子程序创建。

程序代码如下。

```
--声明部分
DECLARE
    str varchar(50) = '';
-- 声明和定义子程序存储过程
PROCEDURE add_str (
    str1 text
)
IS
--异常处理部分
error_message VARCHAR2(50) := '长度超出限制';
--可执行部分
```

```
BEGIN
    str := str || '_' || str1;
EXCEPTION
    WHEN VALUE_ERROR THEN
        RAISE NOTICE '%', error_message;
END add_str;
--调用子程序
BEGIN
    add_str('第一次调用'); -- invocation
    RAISE NOTICE '调用结果为 %', str;
    add_str('第二次调用'); -- invocation
    RAISE NOTICE '调用结果为 %', str;
    add_str('第三次调用'); -- invocation
    RAISE NOTICE '调用结果为 %', str;
END;
```

程序运行结果如下。

```
调用结果为 _第一次调用
调用结果为 _第一次调用_第二次调用
SQL 错误 [22001]: 错误: 对于可变字符类型来说,值太长了(15)
  WHERE: PL/SQL 函数 add_str 的第 12 行的赋值
SQL 语句 "CALL add_str('第三次调用')"
PL/SQL 函数 inline_code_block 的第 23 行的 CALL
```

示例 9.10：在存储过程中应用嵌套子程序。

程序代码如下。

```
--创建带子程序的存储过程,功能为卖出货物后更新商品库存,该存储过程将图 9.1 中的函数作
为嵌套子程序嵌套在了本存储过程中
--头部
--依次定义了货物编码、商店编码、卖出货物数量 3 个参数,数据类型为整数
CREATE OR REPLACE PROCEDURE "updatesupply_totlwhamt2" (
    IN s_goodid BIGINT, IN s_shopid BIGINT, IN s_saleamt INTEGER
) AS
--声明部分
DECLARE
surplus INTEGER;
--将图 9.1 的示例函数创建为子程序,功能为判断剩余库存是否满足卖出数量
--依次定义了货物编码、商店编码、卖出货物数量 3 个参数,数据类型为整数
PROCEDURE _chksupply_totlwhamt (
    s_goodid BIGINT,  s_shopid BIGINT,  saleamt INTEGER
) IS
BEGIN
    --查询指定商店的指定货物的库存数量,并减去传入的卖出货物数量
    SELECT totlwhamt-saleamt
    FROM  "supply"
    WHERE shopid = s_shopid AND goodid = s_goodid   INTO surplus;
```

```
    END _chksupply_totlwhamt;

--存储过程主体执行部分
BEGIN--调用子程序计算货物剩余库存是否满足卖出数量
    _chksupply_totlwhamt(
    s_goodid,
    s_shopid,
    saleamt
);
--如果库存足够,卖出并更新剩余库存
    IF surplus>0 THEN
    UPDATE "supply"
        SET "totlwhamt" = "totlwhamt" -s_saleamt
    WHERE shopid = s_shopid AND goodid = s_goodid;
    END IF;
END;
```

9.3.3 支持的参数

嵌套子程序参数和普通函数、存储过程中的参数说明一样,可以指定参数模式、参数默认值等。嵌套子程序与普通函数实参表示法相同,参数的指定允许以位置参数或命名参数两种格式。具体参见9.2独立子程序模块。

唯一区别是,嵌套子程序的参数可以定义在 PL/SQL 块中的集合类型或自定义类型,而普通函数或存储过程的参数则不可以。

示例 9.11：嵌套子程序参数示例,参数是自定义 record 类型。

程序代码如下。

```
DECLARE
TYPE record_parameter   IS RECORD(totlwhamt
INT,saleamt
INT ); --自定义了名为 record_parameter 的是自定义 record 类型
surplus record_parameter;
PROCEDURE  record_chksupply_totlwhamt( parameter IN OUT record_parameter)
IS
--创建子程序,使用了名为 parameter 的 record_parameter 自定义 record 类型参数
BEGIN
    parameter.totlwhamt:= parameter.totlwhamt-parameter.saleamt;
    raise notice 'parameter.totlwhamt= %', parameter.totlwhamt;
--调用子程序
BEGIN
    surplus.totlwhamt:=99;
    surplus.saleamt:=2;
    record_chksupply_totlwhamt(surplus);
END;
```

程序运行结果如下。

```
NOTICE: parameter.totlwhamt:=97
ANONYMOUS BLOCK
```

9.3.4 调用与变量

嵌套子程序只能在声明它的 PL/SQL 块、函数、存储过程中调用。调用嵌套子程序时，在当前 PL/SQL 块声明中搜索匹配的函数，如果找不到匹配的嵌套子程序，则报错。

示例 9.12：嵌套子程序调用示例。

程序代码如下。

```
DECLARE
    v1 integer;
    v2 integer := 100;
PROCEDURE test(id integer)
AS
    v2 integer := 10;
    v3 integer;
PROCEDURE test(id integer)
AS
v3 integer;
BEGIN
    v3 := 10;
    raise notice 'in the second test function';
END;
BEGIN
    call test(2);
    v3 := v2;
    raise notice 'in the first test function';
END;
BEGIN
    call test(1);
END;
```

程序运行结果如下。

```
NOTICE:
in the second test function
NOTICE:
in the first test function
ANONYMOUS BLOCK
```

把包含嵌套子程序的函数、存储过程、匿名块称为父程序。父程序不可以引用嵌套子程序的变量，但是嵌套子程序可以引用父程序中的变量。

示例 9.13：嵌套子程序引用变量示例。

程序代码如下。

```
CREATE OR REPLACE FUNCTION func1(id integer) RETURN integer AS
```

```
    v1 int := 11;
    v2 int;
FUNCTION func2(id integer) RETURN integer AS
BEGIN
    id := v1;
    raise notice 'v1 = %', v1;
    return id;
END;
BEGIN
    v2:= func2(100);
    return v2;
END;

SELECT func1(1);
```

程序运行结果如下。

```
NOTICE:  v1 = 11
 func1
-------
    11
(1 row)
```

9.4 子程序重载

KingbaseES 数据库允许对子程序进行重载,如果多个不同子程序的形参在个数或数据类型上不同,则可以为多个不同的子程序使用相同的名称。

与 Oracle 不同的是,在 KingbaseES 数据库中,独立子程序、嵌套子程序和包子程序均可以重载,而 Oracle 中的独立子程序不能重载。

示例 9.14:子程序的重载。

功能描述:该示例定义了两个同名的 initialize 子程序,用于初始化不同类型的集合。因为程序中的处理是相同的,所以给它们的命名相同是合乎逻辑的。PL/SQL 通过检查两者的形参确定调用哪个过程。

程序代码如下。

```
DECLARE
  TYPE date_tab_typ IS TABLE OF DATE   INDEX BY INT;
  TYPE num_tab_typ  IS TABLE OF NUMBER INDEX BY INT;

  date_tab    date_tab_typ;
  num_tab     num_tab_typ;

  PROCEDURE initialize (date_tab OUT date_tab_typ, i INTEGER) IS
  BEGIN
    RAISE NOTICE 'Invoked first version';
```

```
    FOR i IN 1..i LOOP
      date_tab(i) := SYSDATE;
    END LOOP;
  END initialize;

  PROCEDURE initialize (num_tab OUT num_tab_typ, i INTEGER) IS
  BEGIN
    RAISE NOTICE 'Invoked second version';
    FOR i IN 1..i LOOP
      num_tab(i) := 1.0;
    END LOOP;
  END initialize;
```

调用相关的子程序,程序代码如下。

```
BEGIN
  initialize(date_tab, 50);
  initialize(num_tab, 100);
END;
```

程序运行结果如下。

```
NOTICE: Invoked first version
NOTICE: Invoked second version
```

9.5 表函数

表函数是多行元组的函数,就像一个关系型表一样,可以在查询语句的 FROM 子句中调用。

实现表函数有以下两种方式。

(1) 结果返回行的集合,可以直接在 FROM 后使用。

(2) 结果返回集合数据类型,使用 TABLE 操作符将结果转换为多行输出。

9.5.1 结果返回行集合

函数的返回值一般是某一类型的值,如 INT、VARCHAR、DATE 等,需要返回结果集时,就要使用 SETOF 语法。

先定义一个嵌套表类型,用于返回函数结果,如下所示。

```
CREATE OR REPLACE TYPE obj_table as OBJECT(
  id INT,
  name VARCHAR2(50)
);
```

示例 9.15：计算从 1 到指定数字的平方值，结果返回行的集合。
程序代码如下。

```
CREATE OR REPLACE FUNCTION f_pipe_kes(s number)
  RETURN SETOF obj_table AS
DECLARE
    item obj_table;
BEGIN
  FOR i IN 1 .. s
  LOOP
      item := obj_table (i,to_char(i * i));
      RETURN NEXT item;
  END LOOP;
  RETURN;
END;
```

调用该函数，程序代码如下。

```
SELECT * FROM f_pipe_kes(5);
```

程序运行结果如下。

```
 id     | name          |
--------+---------------+
 1      | 1             |
 2      | 4             |
 3      | 9             |
 4      | 16            |
 5      | 25            |
```

表函数能够和常规函数一样，接收任何参数，带有游标变量参数的表函数适合用作数据转换函数。

示例 9.16：表函数进行数据转换。
假如需要将顾客的隐私信息（手机号码、邮箱等）单独存放到一个数据结构中，程序代码如下。

```
custid int4,                       --顾客 id
infotype VARCHAR(10),              --信息类型
infodetail VARCHAR(100)            --具体信息
```

即原来表 customers 中的一行数据需要变成两行，操作如下。

```
CREATE OR REPLACE PACKAGE refcur_pkg IS
   TYPE refcur_t IS REF CURSOR RETURN customers%ROWTYPE;
   TYPE outrec_typ IS RECORD (
      custid     int4 ,
      infotype   VARCHAR2(10),
      infodetail VARCHAR2(100));
   --TYPE outrecset IS TABLE OF outrec_typ;
```

```
    FUNCTION f_trans(p refcur_t)
      RETURN SETOF outrec_typ;
END;

CREATE OR REPLACE PACKAGE BODY refcur_pkg IS
    FUNCTION f_trans(p refcur_t)
      RETURN SETOF outrec_typ
    AS
      out_rec outrec_typ;
      in_rec  p%ROWTYPE;
    BEGIN
    LOOP
      FETCH p INTO in_rec;
      EXIT WHEN p%NOTFOUND;
      --email
      out_rec.custid := in_rec.custid;
      out_rec.infotype := 'email';
      out_rec.infodetail := in_rec.email;
      RETURN NEXT out_rec;
      --mobile
      out_rec.custid := in_rec.custid;
      out_rec.infotype := 'mobile';
      out_rec.infodetail := in_rec.mobile;
      RETURN NEXT out_rec;
    END LOOP;
  CLOSE p;
  RETURN;
  END;
END;
```

调用该包内的函数,程序代码如下。

```
DECLARE
  TYPE custcurtype IS REF CURSOR;
  cust_cv custcurtype;
BEGIN
  OPEN cust_cv FOR SELECT * FROM customers LIMIT 2;
  SELECT *
  FROM refcur_pkg.f_trans(cust_cv);
END;
```

程序运行结果如下。

```
custid | infotype | infodetail           |
-------+----------+----------------------+
1      | email    | lin2@qq.com          |
1      | moblie   | 13125948013          |
2      | email    | jiang@foxmail.com    |
2      | mobil    | 13161800215          |
```

9.5.2 结果返回集合数据类型

除了使用 SETOF 返回行集合，表函数还可以返回集合数据类型，它具有如下两种方式。

（1）使用过程变量，对中间过程变量进行赋值，最后一次性返回。

（2）兼容 Oracle 的 PIPELINED 功能，不需要将所有结果都存放到一个集合中，等待函数结束才开始返回数据，只要有一行准备好的数据存放到集合中，函数就可以通过管道将这行数据传输出去，而不是一次返回，也称为管道表函数。

示例 9.17：计算从 1 到指定数字的平方值，使用过程变量。

程序代码如下。

```
CREATE OR REPLACE TYPE t_table IS TABLE OF obj_table;

CREATE OR REPLACE FUNCTION f_normal(s number)
RETURN t_table
AS
rs t_table:= t_table();
BEGIN
    FOR i IN 1..s LOOP
      rs.extend;
        rs(rs.count) := obj_table(rs.count,to_char(rs.count * rs.count));
    END LOOP;
RETURN rs;
END;
```

调用该函数，使用 TABLE 操作符将集合转换为多行输出，程序代码如下。

```
SELECT * FROM TABLE(f_normal(5));
```

程序运行结果如下。

```
id         | name             |
-------+--------------+
1          | 1                |
2          | 4                |
3          | 9                |
4          | 16               |
5          | 25               |
```

示例 9.18：计算从 1 到指定数字的平方值，使用 PIPELINED 功能。

程序代码如下。

```
CREATE OR REPLACE FUNCTION f_pipe(s number)
  RETURN t_table PIPELINED
AS
  v_obj_table obj_table;
BEGIN
```

```
    FOR i IN 1..s LOOP
      v_obj_table :=  obj_table(i,to_char(i * i));
      PIPE ROW(v_obj_table);
    END LOOP;
    RETURN;
END;
```

调用该函数,使用 TABLE 操作符将集合转换为多行输出,程序代码如下。

```
SELECT * FROM TABLE(f_pipe(5));
```

程序运行结果如下。

```
 id     | name           |
--------+----------------+
 1      | 1              |
 2      | 4              |
 3      | 9              |
 4      | 16             |
 5      | 25             |
```

第 10 章

用户自定义对象

KingbaseES 数据库除了具有关系数据库的特性，还具有面向对象的特性。可以在数据库中定义自己的类，即对象类型，也称为用户自定义类型。本章包括以下主要内容。
- 用户自定义对象概述。
- 创建用户自定义对象。
- 在 PL/SQL 中使用自定义对象。
- 与 Oracle 数据库中对象类型的差异。

10.1　用户自定义对象概述

面向对象的编程语言，例如 Java、C♯等，都允许定义类。类定义了属性和方法，属性用于存储对象的状态，方法用于建立对象的行为模型。KingbaseES 数据库中的对象类型与 Java、C♯中的类相似，都可以包含属性和方法。

对象类型是一个用户自定义的复合数据类型，将数据结构和数据操作封装在一起，模拟现实世界中对象的属性和行为。对象类型是对象的模型，对象是对象类型的实例。以对象形式表示现实的事物，使得数据库中的数据更有意义。

10.2　创建用户自定义对象

10.2.1　对象类型

对象类型的定义由对象类型规范和对象类型体构成。对象类型规范声明了对象的公共属性与方法，对象类型体用于实现对象类型规范中声明的方法。如果对象类型规范中没有声明任何方法，则可以没有对象类型体的定义。从面向对象的角度来看，对象类型规范是对象类型的公共接口，而对象类型体是私有实现。

可以使用 CREATE TYPE 语句创建对象类型，语法格式如下。

```
CREATE [ OR REPLACE ] TYPE name [ FORCE ] [ AUTHID { CURRENT_USER | DEFINER } ]
{ AS | IS } OBJECT (
  attribute_name data_type [, ... ]
  [, subprogram_spec [, ... ] ]
)
```

参数说明如表 10.1 所示。

表 10.1　创建对象类型的参数说明

参　　数	描　　述
OR REPLACE	如果对象类型已存在,则重新创建并编译该类型
FORCE	与 OR REPLACE 配合使用的选项参数,当创建一个对象类型,需要替换一个已有类型时,FORCE 选项决定该如何处理被替换类型的依赖关系。含有 FORCE 的替换创建语句会级联删除所有对原类型有依赖的对象(请谨慎使用),不含有 FORCE 的替换创建语句会在检测到原有类型处于被依赖状态时进行提示,并组织新类型成功创建
AUTHID { CURRENT_USER | DEFINER }	AUTHID CURRENT_USER 表示在调用该类型内部方法时,使用调用方法的用户权限来执行。这是默认值。AUTHID DEFINER 指定要用类型的属主的权限来执行该类型的内部方法
AS|IS	创建对象类型时使用关键词 AS 或 IS
attribute_name	对象类型属性的名称
data_type	对象类型属性的数据类型,例如 NUMBER、CHAR、DATE 等,或其他对象类型
subprogram_spec	对象方法声明

在对象类型定义中,可以声明对象方法,包括函数或过程。对象方法在对象类型规范中声明,在对象类型体中实现。对象方法包括以下几种。

(1) MEMBER 方法。

MEMBER 方法通过对象实例调用,用于对实例数据的操作,又称为成员方法或实例方法。在对象类型中可以定义多个 MEMBER 方法。

(2) STATIC 方法。

STATIC 方法是依赖对象类型的方法,与对象实例无关。在对象类型中可以定义多个 STATIC 方法。STATIC 方法不能通过对象实例加以调用,而只能由对象类型调用,调用形式为 type_name.method()。

(3) 构造方法。

构造方法是一个与对象类型同名、返回对象类型实例的函数,用于创建一个对象实例,因此构造方法又称为构造函数。对象类型的构造方法包括系统定义的构造方法和用户定义的构造方法两类。默认情况下,系统隐式地为每个包含属性的对象类型定义一个构造方法,为新建对象的每个属性赋值。用户可以定义多个构造方法,这些方法具有相同的名称,但具有不同的参数个数、顺序、类型,实现构造方法的重载。用户定义构造方法的第一参数为隐式声明的 SELF 参数,表示当前对象。声明构造方法时,该参数是可选的,如果显式指定该参数,参数模式必须为 IN OUT。

与记录、集合等复合数据类型可以在包、PL/SQL 块中定义不同,对象类型只能在数据库中定义,以模式对象的形式独立存储。可以像使用内置数据类型一样使用对象类型,可以声明对象类型的变量,还可以使用对象类型作为表中列的数据类型。

示例 10.1:创建对象类型 t_address。

功能描述:该对象类型可用于表示地址,包含 4 个属性,名称分别为 province、city、

street 和 zip。

程序代码如下。

```
CREATE OR REPLACE TYPE t_address AS OBJECT (
    province    VARCHAR2,           --省份
    city        VARCHAR2,           --城市
    street      VARCHAR2,           --街道
    zip         VARCHAR2            --邮编
);
```

示例 10.2：创建对象类型 t_person。

功能描述：该类型中 address 的类型为已定义的对象类型，还包含对象函数的声明。

程序代码如下。

```
CREATE OR REPLACE TYPE t_person AS OBJECT (
    custid    INTEGER,              --顾客 ID
    custname VARCHAR,               --顾客姓名
    gender    VARCHAR,              --顾客性别
    dob       DATE,                 --出生日期
    email     VARCHAR,              --邮箱
    mobile    VARCHAR ,             --手机号码
    address   t_address,            --地址
    --获取顾客年龄
    MEMBER FUNCTION  get_custage RETURN   NUMBER,
    --获取顾客手机号码(私密显示)
    MEMBER FUNCTION  get_custmobile RETURN   VARCHAR2 ,
    --获取顾客邮箱(私密显示)
    MEMBER FUNCTION  get_custemail RETURN   VARCHAR2 ,
    --静态获取当前时间
    STATIC FUNCTION  get_date RETURN DATE,
    --自定义构造函数
    CONSTRUCTOR FUNCTION t_person(
        p_custid INTEGER,
        p_custname VARCHAR,
        p_gender VARCHAR,
        p_dob DATE,
        p_email VARCHAR,
        p_mobile VARCHAR
    ) RETURN SELF AS RESULT
);
```

在该类型中，除了基本的属性外，还声明了 3 个 MEMBER 方法，分别获取顾客的年龄、手机号码(脱敏处理)和邮箱(脱敏处理)；声明了 1 个 STATIC 方法，获取当前日期；声明两个构造函数用来初始化对象。

由于 t_person 中包含对象方法声明，所以必须为 t_person 创建对象类型体，对象类型体中包含对象方法的实现，通过 CREATE TYPE BODY 语句创建，下面为 t_person 创建对象类型体。

示例 10.3：创建对象类型体。

程序代码如下。

```
CREATE TYPE BODY t_person
  AS
  --获取顾客的年龄
  MEMBER FUNCTION get_custage RETURN NUMBER IS
    v_custage NUMBER;
  BEGIN
    SELECT TRUNC(months_between(sysdate,dob)/12) INTO v_custage FROM dual;
    RETURN v_custage;
  END;
  --获取手机号码(脱敏处理)
  MEMBER FUNCTION  get_custmobile RETURN VARCHAR2 IS
    v_mobile VARCHAR2;
  BEGIN
    v_mobile:=replace(mobile,substr(mobile,4,4),'****');
    RETURN v_mobile;
  END;
  --获取电子邮箱(脱敏处理)
  MEMBER FUNCTION  get_custemail RETURN VARCHAR2  IS
    v_email VARCHAR2(50);
  BEGIN
    v_email:=replace(email,substr(email,4,4),'****');
    RETURN v_email;
  END;
  --获取当前日期
  STATIC FUNCTION get_date RETURN DATE IS
  BEGIN
    RETURN SYSDATE;
  END;
  --自定义构造函数:不包含地址信息
  CONSTRUCTOR FUNCTION t_person(
    p_custid INTEGER,
    p_custname VARCHAR,
    p_gender VARCHAR,
    p_dob DATE,
    p_email VARCHAR,
    p_mobile VARCHAR
  ) RETURN SELF AS RESULT IS
  BEGIN
    SELF.custid:=p_custid;
    SELF.custname:=p_custname;
    SELF.gender:=p_gender;
    SELF.dob:=p_dob;
    SELF.email:=p_email;
```

```
    SELF.mobile:=p_mobile;
    RETURN;
  END;
END;
```

该对象类型体中详细描述了每个对象方法如何实现,相关方法的解释如下。

get_custage 获取顾客年龄,通过 months_ between()函数计算当前日期与出生日期的月份差值,再除以 12,得到顾客年龄。

get_custmobile 和 get_custemail 对手机号码和邮箱进行脱敏处理,例如 123456@qq.com 显示为 123***@qq.com。

t_person 是自定义构造函数,对除了 address(地址)的其他属性赋值。

10.2.2 对象实例

一个对象类型的变量就是对象类型的实例,简称对象。一个对象类型的实例代表一个具体的事物,可以为其属性赋值,并调用其方法。

示例 10.4:创建 t_person 类型的对象实例。

程序代码如下。

```
DECLARE
  ass t_address;
  cust1 t_person;
  cust2 t_person;
  cust3 t_person;
BEGIN
  --使用默认构造函数初始化对象实例
  ass:=t_address('北京市','北京市','朝阳区','100024');
  cust1:=t_person(1,'张三','男','2000-1-1','1234567@139.com','13342212345',
ass);
  --使用自定义构造函数初始化对象实例
  cust2:=t_person(2,'李四','女','1990-1-1','1111111@qq.com','13563199999');
  --使用属性赋值
  cust3.custid:=3 ;
  cust3.custname:='王五';
  cust3.gender:='女';
  cust3.dob:='1894-10-8';
  cust3.email:='wangwu@mail.com';
  cust3.mobile:='18840866666';
  cust3.address:=ass;

  RAISE NOTICE '%',cust1;
  RAISE NOTICE '姓名:%,年龄:%,手机号码:%',cust2.custname,cust2.get_custage(),
cust2.get_custmobile();
  RAISE NOTICE '姓名:%,地址:%',cust3.custname,(cust3.address.province||cust3.
address.city||cust3.address.street);
END;
```

程序运行结果如下。

```
NOTICE:(1,张三,男,"2000-01-01 00:00:00",1234567@139.com,13342212345,"(北京市,北
京市,朝阳区,100024)")
NOTICE:姓名:李四,年龄:33,手机号码:135****9999
NOTICE:姓名:王五,地址:北京市北京市朝阳区
```

10.3 在 PL/SQL 中使用自定义对象

在 PL/SQL 中可以创建和操作对象。只要在模式中定义了对象类型，就可以在任意 PL/SQL 块、子程序或包中引用它来声明对象。也可以使用对象类型作为变量、形参或函数返回值的数据类型。

10.3.1 定义对象

在 PL/SQL 中，可以在使用内置类型（NUMBER 或 CHAR 等）的地方使用对象类型来声明一个对象。

示例 10.5：声明了 t_address 类型的对象 ass，然后调用构造函数初始化对象。

程序代码如下。

```
DECLARE
  ass t_address;
BEGIN
  ass:= t_address('山东省','文登市','龙山街道','264400');
  RAISE NOTICE '%,%,%',ass.province,ass.city,ass.street;
END;
```

还可以将对象类型作为函数和过程的形参，把对象从一个子程序传递给另一个子程序。程序代码如下。

```
DECLARE
  ...
  PROCEDURE init_address(new_address IN OUT t_address) IS ...
```

还可以将对象类型作为函数的返回类型来使用。程序代码如下。

```
DECLARE
  ...
  FUNCTION get_address(zip IN VARCHAR2) RETURN t_address IS ...
```

10.3.2 初始化对象

如果声明对象后不调用构造函数初始化对象，则会被自动赋空值，不仅是它的属性，而且对象本身也为空。

示例 10.6：声明对象，不进行初始化。

程序代码如下。

```
DECLARE
  ass t_address;
BEGIN
  IF ass IS NULL THEN
    RAISE NOTICE '对象为 NULL';
  END IF ;
END;
```

运行后可以看到输出了"对象为 NULL"的提示。在 PL/SQL 中，如果为一个未初始化的对象属性赋值，会引发预定义异常，一个空对象不能等于另一个对象，保持良好的编程习惯就是在声明的时候为对象初始化。

10.3.3 调用构造函数

调用构造函数时，会将对象中需要初始化的属性赋予初始值。如果调用默认的构造函数，则需要为每一个属性指定一个初始值。

只要能够调用普通函数的地方，就可以调用构造函数。和所有的函数一样，构造函数也可以作为表达式的一部分被调用。

示例 10.7：构造函数作为表达式的一部分被调用。

程序代码如下。

```
DECLARE
  ass VARCHAR2 ;
  FUNCTION show_address(_ass t_address)
    RETURN VARCHAR2 IS
  BEGIN
    RETURN _ass.province||_ass.city||_ass.street;
  END ;
BEGIN
  ass:=show_address(t_address('山东省','文登市','龙山街道','264400'));
  RAISE NOTICE '%',ass;
END;
```

程序运行结果如下。

NOTICE：山东省文登市龙山街道

10.3.4 调用 MEMBER 方法和 STATIC 方法

MEMBER 方法通过对象实例调用，调用方法为 object.method()。

STATIC 方法是依赖于对象类型的方法，调用方法为 type_name.method()。

示例 10.8：调用 MEMBER 获取部分顾客的年龄和邮箱，调用 STATIC 方法获取当前日期。

程序代码如下。

```
DECLARE
  customer t_person;
  v_today DATE;
  v_name VARCHAR2;
  v_age VARCHAR2;
  v_email VARCHAR2;
BEGIN
  SELECT t_person.get_date() INTO v_today;
  customer:=t_person(1,'张三','男','2000-1-1','123456@qq.com','12312312345');
  SELECT customer.custname,customer.get_custage(),customer.get_custemail()
  INTO v_name,v_age,v_email;
  RAISE NOTICE '现在是%,%年龄:%,邮箱:%',v_today,v_name,v_age,v_email;
END;
```

程序运行结果如下。

```
NOTICE:现在是 2023-02-14 17:02:10,张三年龄:23,邮箱:123****qq.com
```

10.3.5 对象表的 DML 操作

可以使用对象类型定义表中的单个列,存储在此列中的对象称为列对象;也可以使用对象类型定义表中的一整行,这张表称为对象表。在 PL/SQL 中,可以直接对对象表进行 DML 操作。

示例 10.9:创建存储类型为 t_person 的对象表,并对其进行 DML 操作。

程序代码如下。

```
DROP TABLE IF EXISTS persons;
CREATE TABLE persons OF t_person;

BEGIN
  --向对象表插入行,通过构造函数提供属性值
  INSERT INTO persons
  VALUES (t_person(1,'张三','男','2000-1-1', '1234567@139.com','13342212345',t_address('山东省','文登市','龙山街道','264400')));
  --忽略构造函数,像在关系表中一样提供列值
  INSERT INTO persons(
    custid,custname,gender,dob,email,mobile
  ) VALUES(
    2,'李四','女','1990-10-25','1111111@qq.com','13563199999'
  );
  --从对象表中检索这些行
  SELECT * FROM persons;
  --修改李四的地址
  UPDATE persons
  SET address=t_address('北京市','北京市','朝阳区','100024')
  WHERE custid=2;
```

```
  --根据 address 列对象中的 province 属性检索
  SELECT p.custname,p.get_custage(),(p.address).city,(p.address).street FROM persons p
  WHERE (p.address).province='山东省';
  --删除表中的对象
  DELETE FROM persons
  WHERE custid=1;
END;
```

程序运行结果如下。

```
custid | custname | gender | dob        | email           | mobile      | address
-------+----------+--------+------------+-----------------+-------------+-----------
1      | 张三     | 男     | 2000-01-01 | 12345@139.com   | 13342212345 | [山东省,文登市,龙山街道,264400]
2      | 李四     | 女     | 1990-10-25 | 11111@qq.com    | 13342212345 | [ , , ]
```

```
custname | t_person.get_custage | city    | street
---------+----------------------+---------+---------
张三     | 23                   | 文登市  | 龙山街道
```

10.4 与 Oracle 数据库中对象类型的差异

在 KingbaseES 数据库当前版本（V9）的对象类型中，可以声明属性成员和对象方法，但不支持继承和重载。而 Oracle 数据库中的对象类型支持继承和重载。如果在 Oracle 中使用了对象类型的继承特性，可以在迁移至 KingbaseES 数据库中时采取折中方案。

示例 10.10：手动模拟 Oracle 数据库中对象类型继承和重载的特性。

下列代码是在 Oracle 数据库中声明一个名为 rectangle 的对象类型，包含两个属性和两个成员方法。

```
  --创建基础类型
CREATE OR REPLACE TYPE rectangle FORCE AS OBJECT(
  length NUMBER,
  width NUMBER,
  MEMBER FUNCTION enlarge(inc number) RETURN rectangle,
  NOT FINAL MEMBER PROCEDURE display
) NOT FINAL;
  --创建基础类型的对象类型体
CREATE OR REPLACE TYPE BODY rectangle AS
  MEMBER FUNCTION enlarge(inc number) return rectangle IS
  BEGIN
    RETURN rectangle(self.length + inc, self.width + inc);
  END;

  MEMBER PROCEDURE display IS
```

```
  BEGIN
    dbms_output.put_line('Length: '|| length);
    dbms_output.put_line('Width: '|| width);
  END;
END;
```

随后创建子类型 tabletop，继承自基础类型 rectangle。子类型中除了继承父类型的属性和方法，还声明了 material 属性，对 display 方法进行了重载。

```
--创建子类型
CREATE OR REPLACE TYPE tabletop UNDER rectangle(
  material VARCHAR2(20),
  OVERRIDING MEMBER PROCEDURE display    --重载:重新编写基类的方法
)
--创建子类型的对象类型体
CREATE OR REPLACE TYPE BODY tabletop AS
  OVERRIDING MEMBER PROCEDURE display IS
  BEGIN
    dbms_output.put_line('Length: '|| length);
    dbms_output.put_line('Width: '|| width);
    dbms_output.put_line('Material: '|| material);
  END display;
END;
```

在 KingbaseES 数据库中，需要手动模拟对象类型继承的过程，将父类型的属性和方法整合至子类型，可以改动上述代码如下。

```
--创建对象类型 tabletop1,整合父类型和子类型的属性和方法
CREATE OR REPLACE TYPE tabletop1 as OBJECT(
  length NUMBER,
  width NUMBER,
  material VARCHAR2(20),
  MEMBER PROCEDURE display --重载:重新编写基类的函数
);
-- 为对象类型 tabletop 创建对象类型体
CREATE OR REPLACE TYPE BODY tabletop1 AS
  MEMBER PROCEDURE display IS
  BEGIN
    RAISE NOTICE 'Length:% ',length;
    RAISE NOTICE 'Width: %',width;
    RAISE NOTICE 'Material: %',material;
  END;
END;
```

测试改写后的运行结果如下。

```
DECLARE
  t1 tabletop1;
  t2 tabletop1;
```

```
BEGIN
  t1:= tabletop1(20, 10, 'Wood');
  t2:= tabletop1(50, 30, 'Steel');
  t1.display();
  t2.display();
END;
```

可以看到模拟 Oracle 数据库中对象类型继承和重载后的情景,结果如下。

```
NOTICE:Length:20
NOTICE:Width: 10
NOTICE:Material: Wood
NOTICE:Length:50
NOTICE:Width: 30
NOTICE:Material: Steel
```

在 Oracle 数据库的对象类型中,还有一类特殊的成员方法,即比较方法(MAP 和 ORDER),主要是为了告诉 Oracle,比较该类型的两个对象时应该怎么做。

MAP 方法可以返回对象值的一个映射,ORDER 方法可以比较两个对象,并且返回一个标志值,指明两个对象的相对顺序。Oracle 在需要的时候会自动调用比较方法,比如相等比较或排序时。

在 KingbaseES 数据库中,可以将比较方法定义为普通的 MEMBER 方法,需要时手动调用。

示例 10.11:声明 MEMBER 方法返回对象值的映射,作为排序的依据。

程序代码如下。

```
--创建对象类型
CREATE OR REPLACE TYPE t_man AS OBJECT(
  name VARCHAR2,                              --姓名
  birthday DATE ,                             --出生日期
  MEMBER FUNCTION get_age() RETURN NUMBER     --返回出生日期->年龄的映射
);
--创建对象类型体,实现成员方法
CREATE OR REPLACE TYPE BODY t_man as
  MEMBER FUNCTION get_age() RETURN NUMBER AS
  BEGIN
    RETURN DATE_PART('year',AGE(birthday));
  END;
END;
--在比较对象时调用成员函数,将对象间的比较转换为对象值的比较
DECLARE
  man1 t_man:=t_man('Tom','2000-1-1');
  man2 t_man:=t_man('Jane','1990-1-1');
BEGIN
  IF man1.get_age()>man2.get_age() THEN
    RAISE NOTICE '%比 % 大',man1.name,man2.name;
  ELSE
    RAISE NOTICE '%比 % 大',man2.name,man1.name;
  END IF;
END;
```

第 11 章

用户自定义聚集函数

聚集函数往往用作对查询结果进行分组,然后对每个组返回一个结果。而自定义聚集函数通常基于封装需要,用于改善代码的可读性和基于技术细节的保密要求。本章包括以下主要内容。

- 用户自定义聚集函数概述。
- 用户自定义聚集函数的运用。
- KingbaseES 与 Oracle 中创建聚集函数的差异。

11.1 用户自定义聚集函数概述

11.1.1 聚集函数

聚集函数也叫分组函数,是对多行进行计算的一种函数,例如 SUM、COUNT、AVG、MAX、MIN 等。

聚集函数执行过程中,聚集函数对每一条输入的元组进行处理,运行方式可以看作是对输入值进行处理并完成状态转移的过程,如图 11.1 所示。

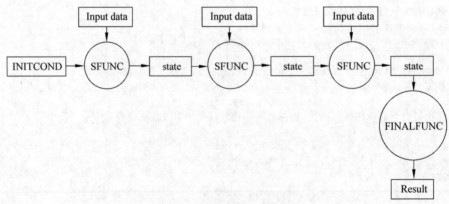

图 11.1 聚集函数运行过程图

聚集函数的实现一般由以下 3 个部分协同完成。

(1) 初始状态值 INITCOND：初始状态的具体值。

(2) 状态转移函数 SFUNC：根据当前输入值 cur_input、状态 cur_state 决定下一次状态的值。

（3）结束收尾函数 FINALFUNC：处理最后一步的状态转换。

一个简单的聚集函数由一个或多个普通函数组成：一个状态转移函数 sfunc 和一个可选的最终计算函数 ffunc。其使用形式如下。

```
sfunc( internal-state, next-data-values ) ---> next-internal-state
ffunc( internal-state ) ---> aggregate-value
```

其中，状态转移函数被每个输入行调用，传入内部聚集状态和当前行的值。

KingbaseES 创建一个数据类型 stype 的临时变量来保持聚集的当前内部状态。对每一个输入行，聚集参数值被计算，并且状态转移函数被调用，用当前状态值和新参数值计算一个新的内部状态值。当所有行均被处理完成后，调用一次最终函数，计算该聚集的返回值。如果没有最终函数，则返回最终的状态值。

11.1.2 创建用户自定义聚集函数

1. 相关语句

CREATE AGGREGATE 是 KingbaseES 的语言扩展。SQL 标准没有提供用户定义的聚集函数。

CREATE AGGREGATE 定义一个新的聚集函数。

CREATE OR REPLACE AGGREGATE 将定义新的聚集函数或替换现有定义。

2. 常用语法

语法格式如下。

```
CREATE [ OR REPLACE ] AGGREGATE name ( [ argmode ] [ argname ] arg_data_type
[ , ... ] ) (
    SFUNC = sfunc,
    STYPE = state_data_type
    [ , FINALFUNC = ffunc ]
    [ , INITCOND = initial_condition ]
    [ , COMBINEFUNC = combinefunc ]
)
```

3. 相关语义

语义描述如下。

name 是要创建的聚集函数的名称(可以是模式限定的)。

argmode 是一个参数的模式：IN 或者 VARIADIC(聚集函数不支持 OUT 参数)。如果忽略，默认值是 IN。只有最后一个参数可以被标记为 VARIADIC。

argname 是参数的名称。如果被忽略，则该参数没有名称。

arg_data_type 是聚集函数操作的一个输入数据类型。要创建一个零参数的聚集函数，可以用 * 来替代参数说明的列表，如 count(*)。

sfunc 为每一个输入行调用的状态转移函数名。对于一个正常的 N-参数聚集函数，sfunc 必须接收 $N+1$ 个参数，第一个参数的类型是 state_data_type，而其余的参数匹配该聚集被声明的输入数据类型，返回一个类型为 state_data_type 的值。该函数采用当前的状态值以及当前的输入数据值，并且返回下一个状态值。

　　state_data_type 是聚集的状态值的数据类型。

　　ffunc 是最终函数的名称，该函数在所有输入行遍历之后被调用，以此计算聚集的结果。对于一个常规聚集，该函数须只接收一个类型为 state_data_type 的单一参数。该聚集的返回数据类型被定义为这个函数的返回类型。如果没有指定 ffunc，则结束状态值被用作聚集的结果，并且返回值类型为 state_data_type。

　　对于有序集（包括假想集）聚集，最终函数不仅接收最终状态值，还会接收所有直接参数的值。

　　如果指定了 FINALFUNC_EXTRA，则除了最终状态值和任何直接参数外，最终函数还接收额外的对应于该聚集的常规（聚集）参数的 NULL 值。主要用于定义一个多态聚集时，允许输出的聚集结果类型。

　　initial_condition 是初始状态值，为常量，数据类型为 state_data_type。如果没有指定，则状态值从空值开始。

　　combinefunc 可以被有选择地指定，以允许聚集函数支持部分聚集。如果提供这个函数，combinefunc 必须组合两个 state_data_type 值，每一个都包含在输入值某个子集上的聚集结果，将产生一个新的 state_data_type，用以表示在两个输入集上的聚集结果。可以把这个函数看作一个 sfunc，后者是在一个个体输入行上操作，并且把它加到运行聚集状态上，而该函数则是把另一个聚集状态加到运行状态上。

　　combinefunc 必须被声明为有两个 state_data_type 参数，并且返回一个 state_data_type 值。这个函数可以有选择性地被标记为 strict。在被标记的情况下，当任何一个输入状态为空时，将不会调用该函数，而是把另一个状态当作正确的结果。

　　对于 state_data_type 为 internal 的聚集函数，combinefunc 不能为 strict。在这种情况下，combinefunc 必须正确处理空状态，并且被返回的状态能够恰当地存储在聚集内存上下文中。

11.2　用户自定义聚集函数的运用

　　在实际业务中，对于订单运送事务，如何获取快递运费的最大值？通常，快递运费的定价标准有两种方式：使用重量定价或结合体积定价。而同一订单可以包含不同的商品，因此，简单的聚集函数，如 SUM 并不能满足需求，可以自定义一个类似聚集函数 SUM 的功能函数，以满足需求。

11.2.1　场景数据

　　先查询一个包含多个商品的订单，得到订单编号。以编号"236995464477"为例，程序代码如下：

```
SELECT A."ordid"  , C."goodid",C."grossweight",C."v_size"
FROM Orders A LEFT JOIN
```

```
      Lineitems B ON A."ordid" =B."ordid" LEFT JOIN
      goods C ON B."goodid" =C."goodid"
WHERE A.ordid='236995464477'
GROUP BY A.ordid, C."goodid"
```

同一订单查询结果如下。

ordid	goodid	grossweight	v_size	ordid
236995464477	100232901206	0.55	3.4	236995464477
236995464477	68327953786	0.25	2.3	236995464477

查询获得一个包含多个商品的订单后插入测试数据，请确保订单商品的毛重和体积有数据。然后确定一个运费标准：起步价为 0.2 元，以重量计算 0.2 元/单位，以体积计算 0.1 元/单位，程序代码如下。

```
--插入一个体积字段
alter table "goods" add V_size double null;
--插入测试数据
update  goods set V_size='3.4' where goodid='100232901206'
update  goods set V_size='2.3' where goodid='68327953786 '
```

11.2.2 创建用户自定义聚集函数

1. 准备状态转移函数 sfunc

内部聚集状态包含：当前商品运费总价格加上按重量计算商品总价格（或者当前商品运费总价格加上按体积计算商品总价格），和当前最大商品运费价格。weight_price 作为以毛重为标准的单位价格，size_price 作为以体积为标准的单位价格，可自行定义。

通过对 state_price 和当前商品运费价格求和，获得新的当前运费总和，程序代码如下。

```
--状态转换函数
CREATE or replace function cal_transform_price_sfunc(state_price double,weight
_price double ,grossweight double ,size double ,size_price double  )
/*声明部分,定义变量、数据类型、异常、局部子程序等*/
return double
as
declare
  weight_price_sum double ;          --按重量计算商品总价格
  size_price_sum double ;            --按体积计算商品总价格
/*执行部分,实现块的功能*/
begin
  if state_price is null
  state_price:=0;
  weight_price_sum := state_price + grossweight * weight_price;
                                    --求和获得当前按重量计算商品总价格的总和
  size_price_sum := state_price + size * size_price;
  --比较大小,找到新的最大值
```

```
    if weight_price_sum < size_price_sum then
      state_price := size_price_sum;
    else
      state_price := weight_price_sum;
    end if;
    return   state_price;
end;
```

2. 准备最终函数 ffunc

对应状态转移函数，完成第一步后，回归主题，定义的聚集函数内部状态是 double 类型，输出是 double 类型，还需一个终止函数接收聚集内部状态，并转换为 double 类型，程序代码如下。

```
--最终函数
CREATE function cal_transform_price_finalfunc( state_price double)
returns double
immutable
as
begin
  return   state_price;
end;
```

3. 创建函数

准备好以上两个自定义函数后，就可以创建自定义聚集函数。参考 11.1.2 节的语法写出创建语句，包括状态转换函数、内部聚集状态类型以及终止函数，程序代码如下。

```
--创建函数
drop aggregate if exists cal_transform_price(double,double,double,double);
CREATE aggregate cal_transform_price (double,double,double,double)
(
    sfunc = cal_transform_price_sfunc,--为每一个输入行调用的状态转移函数名
    stype = double,
    initcond=0.2,
    finalfunc = cal_transform_price_finalfunc
);
```

11.2.3 用户自定义聚集函数的使用

自定义聚集函数可以像内置聚集函数一样使用，用在 SELECT、ORDER BY 和 HAVING 语句中。

示例 11.1：创建自定义聚集函数。

功能描述：根据自定义的价格，使用自定义聚集函数查询运费。确定一个运费标准：起步价为 0.2 元，按重量计算为 0.2 元/单位，按体积计算为 0.1 元/单位。

程序代码如下。

```
DECLARE
  weight_price double :=0.2;
  size_price double :=0.1;
BEGIN
  SELECT A.ordid
    ,cal_transform_price(weight_price,C.grossweight,C.v_size  ,size_price)
  FROM Orders A LEFT JOIN
    Lineitems B ON A.ordid =B.ordid LEFT JOIN
      goods C ON B.goodid =C.goodid
  WHERE A.ordid='236995464477'
  GROUP BY A.ordid;
END
```

程序运行结果如下。

```
ordid              | cal_transform_price |
-------------------+---------------------+
236995464477       | 0.77                |
```

11.2.4　查看用户自定义聚集函数信息

1. 查找所有自定义聚集函数

程序代码如下。

```
SELECT DISTINCT(proname) FROM pg_proc WHERE prokink='a';
```

查询结果如下。

```
proname                         |
--------------------------------+
websearch_to_tsquery            |
pg_dependencies_send            |
time_hash_extended              |
time_hash_extended              |
jsonpath_send                   |
pg_ndistinct_out                |
path_recv                       |
```

2. 查询所有 agg 开头的聚集函数

程序代码如下。

```
SELECT * FROM pg_proc WHERE proname like 'agg%';
```

11.3 KingbaseES 与 Oracle 中创建聚集函数的差异

在 Oracle 中，为了增强扩展性，提供了一个自定义聚集函数接口 ODCIAggregate()，通过实现 ODCIAggregate routines 创建自定义聚集函数。在 Oracle 中实现用户自定义函数，每个函数需要实现 4 个 ODCIAggregate 接口函数，这些函数定义了任何一个聚集函数内部需要实现的操作，分别是初始化（initialization）、迭代（iteration）、合并（merging）和终止（termination）。通过定义一个对象类型（object type），在该类型内部实现 ODCIAggregate 接口函数（routines），可以用任何一种 Oracle 支持的语言实现这些接口函数，如 C/C++、JAVA、PL/SQL 等。定义了对象类型，并且相应的接口函数也都在该对象类型体内部实现，就可以创建自定义聚集函数。表 11.1 是 4 个 ODCIAggregate 接口函数的功能描述。

表 11.1 ODCIAggregate 接口函数的功能描述

函 数	描 述
ODCIAggregateInitialize	Oracle 调用此函数来初始化用户定义聚合计算。初始化的聚合上下文作为对象类型实例传回 Oracle
ODCIAggregateIterate	Oracle 反复调用此函数。每次调用时都会传递一个新值（或一组新值）作为输入。当前的聚合上下文也被传入。函数处理新值，并将更新的聚合上下文返回 Oracle。NULL 为基础组中的每个非值调用此函数（NULL 值在聚合期间被忽略，并且不传递给函数）
ODCIAggregateMerge	Oracle 调用此函数来组合两个聚合上下文。该函数将两个上下文作为输入，将它们组合起来，然后返回一个聚合上下文
ODCIAggregateTerminate	Oracle 调用此函数作为聚合的最后一步。该函数将聚合上下文作为输入，并返回结果聚合值

以创建字符串连接聚集函数 STRAGG(string_agg)为例，在 Oracle 中创建聚集函数的方式如下。

```
CREATE OR REPLACE TYPE STRING_AGG_TYPE as object (
  total varchar2(4000),
  static FUNCTION ODCIAggregateInitialize(sctx IN OUT string_agg_type) return number,
  member FUNCTION ODCIAggregateIterate(
    self IN OUT string_agg_type,
    value IN varchar2
  ) RETURN number,
  member FUNCTION ODCIAggregateTerminate(
    self IN string_agg_type,
    returnValue OUT varchar2,
    flags IN number
  ) RETURN number,
  member FUNCTION ODCIAggregateMerge(
    self IN OUT string_agg_type,
    ctx2 IN string_agg_type
  ) RETURN number
```

```
);
CREATE OR REPLACE TYPE BODY STRING _ AGG _ TYPE is static FUNCTION
ODCIAggregateInitialize(sctx IN OUT string_agg_type) return number is
    BEGIN
    sctx:= string_agg_type(null);
        RETURN ODCIConst.Success;
    END;
    member FUNCTION ODCIAggregateIterate(
      self IN OUT string_agg_type,
      value IN varchar2
    ) RETURN number is
    BEGIN
    self.total:= self.total || ',' || value;
        RETURN ODCIConst.Success;
    END;
    member FUNCTION ODCIAggregateTerminate(
      self IN string_agg_type,
      returnValue OUT varchar2,
      flags IN number
    ) RETURN number is
    BEGIN
    returnValue:= ltrim(self.total, ',');
        RETURN ODCIConst.Success;
    END;
    member FUNCTION ODCIAggregateMerge(
      self IN OUT string_agg_type,
      ctx2 IN string_agg_type
    ) RETURN number is
    BEGIN
      self.total:= self.total || ctx2.total;
      RETURN ODCIConst.Success;
    END;
END;
CREATE OR REPLACE FUNCTION STRAGG (input varchar2) RETURN varchar2 PARALLEL _
ENABLE AGGREGATE USING string_agg_type;
```

KingbaseES 对 Oracle 的聚集函数加以改写转换,实现聚集函数的创建。

KingbaseES 对应的改写参考方式如下。

ODCIAggregateIterate 的转换要点说明如下。

(1) 函数参数需要转换,将 Oralce self 的第 1 个参数改为 state_data_type,第 2 个参数为现有参数类型,参数模式都为 in 模式。

(2) 返回值需要修改为 number-> state_data_type。

```
CREATE OR REPLACE FUNCTION ODCIAggregateIterate(self varchar2, varchar2)
return varchar2
AS
```

```
BEGIN
  self || ',' || $2;
  return self;
END;
```

ODCIAggregateTerminate 的转换要点说明如下。

(1) 函数参数需要转换,将第一个参数 Oralce self 改为 state_data_type,参数模式都为 in 模式。

(2) 返回值需要修改为 number-> state_data_type。

(3) PL/SQL 块中将对原 self 的引用修改为参数 1,并修改返回语句为 out 参数值。

```
CREATE OR REPLACE FUNCTION ODCIAggregateTerminate(varchar2)
return varchar2
AS
BEGIN
  RETURN ltrim($1, ',');
END;
```

ODCIAggregateMerge 的转换要点说明如下。

(1) 函数参数需要转换,将 Oralce self 的第 1 个参数改为 state_data_type,第 2 个参数为现有参数类型,参数模式都为 in 模式。

(2) 返回值需要修改为 number-> state_data_type。

(3) PL/SQL 块中将对原 self 的引用修改为参数 1,修改返回语句为最终求出的参数值。

```
CREATE OR REPLACE FUNCTION ODCIAggregateMerge(varchar2, varchar2)
return varchar2
AS
BEGIN
  return $1 || $2;
END;
```

结合示例 11.1 中创建的用户自定义聚集函数语法创建如下函数,并应用。

示例 11.2:字符串连接聚集函数。

程序代码如下。

```
CREATE AGGREGATE STRAGG(varchar2)
(
  sfunc = ODCIAggregateIterate,        --对应方法 ODCIAggregateIterate
  stype = varchar2,                    --对应的类型
  finalfunc = ODCIAggregateTerminate,  --对应 ODCIAggregateTerminate
  combinefunc = ODCIAggregateMerge,    --对应 ODCIAggregateMerge
  PARALLEL = SAFE,                     --对应 属性 PARALLEL
  --INITCOND = ''                      --对应 ODCIAggregateInitialize,不指定默认为 NULL
);
```

```
\set SQLTERM ;
CREATE table tbl_agg(i varchar2);
INSERT INTO tbl_agg values('aa');
INSERT INTO tbl_agg values('bb');
INSERT INTO tbl_agg values('cc');
INSERT INTO tbl_agg values('dd');

SELECT STRAGG(i) FROM tbl_agg;
```

第 12 章

程 序 包

包是数据库中相关 PL/SQL 元素的集合,开发业务包是数据库开发的重要环节。包又分为用户自定义包与系统内置包,本章将主要介绍用户自定义包,包括以下内容:
- 程序包概述。
- 创建程序包。
- 程序包的使用。

 ## 12.1　程序包概述

12.1.1　包的概念

程序包(PACKAGE,简称包)是一组相关过程、函数、变量、常量和游标等 PL/SQL 程序设计元素的组合,作为一个完整的单元存储在数据库中,用名称来标识包。

与高级语言中的类相同,包中的程序元素也分为公有元素和私有元素两种,二者的区别是允许访问的程序范围不同,即作用域不同。公有元素不仅可以被包中的函数和过程调用,也可以被包外的 PL/SQL 程序访问,而私有元素只能被包内的函数和过程访问。

当然,不包含在程序包中的过程、函数是独立存在的。一般是先编写独立的过程与函数,待其较为完善或经过充分验证无误后,再按逻辑相关性组织为程序包。

12.1.2　包的优点

包在软件工程领域有着悠久的历史,利用包可以实现代码复用,提高程序的可靠性、可维护性以及应用程序的开发效率。在团队开发大型系统过程中,包往往很有效。

(1) 模块化。

通过使用包可以把逻辑上相关的变量、函数等封装成为一个 PL/SQL 模块。每个包都易于理解,包与包之间的接口也简单明了,定义也很方便,有利于应用开发。

(2) 应用程序设计简单。

设计一个应用程序时,最初需要的是包规范的接口信息。可以先编写包规范代码,并编译,而不需要定义包体;包体可以直到需要完成应用程序时再定义。

(3) 隐藏实现细节。

可以共享包规范中的接口信息,隐藏包体中的实现细节。隐藏包体中的实现细节,可以改变包体的具体实现,而不影响包规范中的应用接口信息,应用开发者不能基于包体的实现细节进行开发。如果修改包体的实现,对开发者没有影响。

(4) 功能增强。

包中的变量和游标在一个会话中会一直存在,可以被当前会话中的包函数和包存储过程共享,可以在普通的存储过程和函数中使用包中全局的变量、游标、包的存储过程和包的函数。可以在整个事务中保持对数据的访问,而不必存储在数据库中。

(5) 高效性。

第一次调用包中的子程序时,KingbaseES 数据库加载整个包到内存中。后续对包中其他子程序的调用都在内存中进行,而不需要磁盘 I/O 操作。使用包避免了层叠依赖性以及不必要的重新编译。例如,修改了包中函数的实现,此时并不需要重新编译调用该函数的其他子程序,因为这些子程序仅仅依赖于包规范,而不依赖于包体。

(6) 简化授权。

可以将包的整个权限授予角色,而不需要将包中每个对象的权限授予角色。

12.1.3 系统内置包

KingbaseES 数据库提供了一些系统内置包,开发应用程序时可以直接使用系统内置包中的过程与函数,这既简化了应用开发的复杂性,又提高了程序的运行性能。表 12.1 对系统内置包进行了简单描述,更详细的语法和功能请参考 KingbaseES PL/SQL 过程语言参考手册。

表 12.1 系统内置包及其功能描述

包 名 称	描 述
STANDARD	定义和扩展 PL/SQL 语言环境
DBMS_OUTPUT	处理 PL/SQL 块和子程序输出调试信息
DBMS_RANDOM	提供随机数生成器
UTL_FILE	用 PL/SQL 程序来读写操作系统文本文件
DBMS_DDL	提供了在 PL/SQL 块中执行 DDL 语句的方法以及一些 DDL 的特殊管理方法
DBMS_SESSION	访问或者设置会话信息
DBMS_ALERT	生成并传递数据库预警信息
DBMS_PIPE	同一数据库实例下的不同会话之间进行通信
DBMS_JOB	用于安排和管理数据库中的任务队列
DBMS_LOB	用于访问和操纵 LOB 对象中的全部或部分数据
DBMS_SQL	允许用户使用动态 SQL 语句,构造和执行任意 DML 或 DDL 语句

12.2 创建程序包

12.2.1 包的组成

包由两部分组成:包规范和包体。

包规范是包的说明部分,是对外的操作接口(类似 Java 接口),对应用是可见的。

包体是包的代码和实现部分(类似 Java 实现类),对应用来说是不可见的。

1. 包规范

包规范是指定义包中可以被外部访问的部分,在包规范中声明的内容可以从应用程序和包的任何地方访问。包规范中可以声明类型、变量、常量、函数、存储过程、游标、异常等。函数和存储过程不能在包规范中实现,包规范中变量之间、变量与函数之间不允许重名,函数和存储过程之间可以同名,视为重载。要引用可见域内包中的公共项,需要使用包名称加以限定。每个公共项的声明都包含了使用该公共项所需的所有信息。例如,假设包规范以下面的方式声明函数 func。

```
FUNCTION func (n INTEGER) RETURN INTEGER;         --returns n
```

该声明表明,函数 func 需要一个 INTEGER 类型参数,并返回一个 INTEGER 类型的值。调用者必须知道应该使用什么值调用函数 func,而不需要知道 func 是如何实现的。

包中的公共项包括以下内容。

(1) 可被多个子程序使用的类型、变量、常量、子程序、游标和异常。

包规范中定义的类型可以是 PL/SQL 用户自定义的子类型或 PL/SQL 复合类型。

(2) 可用于独立子程序参数的关联数组类型。

不能在模式中定义关联数组类型。因此,需要将关联数组变量作为参数传递给独立子程序时,必须在包规范中声明关联数组类型。这样可以使被调用的子程序(声明该类型的参数)和调用子程序或匿名块(声明该类型的变量)都能使用该类型。

(3) 在同一会话中的子程序间必须保持可用的变量。

(4) 用于读写公共变量的子程序。

提供这些子程序,以防止包用户直接读写公共变量。

(5) 相互调用的子程序。

不必像关心相互调用的独立子程序一样考虑包中子程序的编译顺序。

(6) 重载的子程序。

重载的子程序是同一子程序的变体。也就是说,它们具有相同的名称,但参数不同。

2. 包体

只有在包规范已经创建的情况下才可以创建包体。包体和包规范必须在同一个模式中。

在包规范中声明并在包体中定义的子程序是可以被包外部引用的公共项。包体还可以声明和定义不能从包外部引用但可用于包的内部的私有项。

包体还可以有一个初始化部分,用于初始化公共变量,并执行一些一次性逻辑。初始化部分仅在第一次引用包时运行。初始化部分可以包括异常处理程序。

可以在未更改包规范中公共项的情况下更改包体。

包体主要用来实现包规范中的函数和存储过程,以及完成变量的初始化等工作。包体严格依赖于包规范,具体地说:

(1) 包规范先于包体创建。

(2)包体不能脱离包规范单独存在,但包规范可以独立存在,如果包规范中没有声明任何子程序,可以不创建包体。

(3)一旦创建包体,就要定义包规范中的所有函数和存储过程。

(4)允许创建不包含任何函数或存储过程的包规范和包体。

在包规范中,可以通过AUTHID短语来确定包中的游标或子程序在运行时是使用定义者权限(默认),还是调用者权限,以及确定是使用对定义者模式对象还是调用者模式对象进行无限制引用。

示例12.1:匹配包规范和包体。

功能描述:在本例中,相应的子程序声明和定义的参数名不匹配;因此,PL/SQL会引发异常。

程序代码如下。

```
CREATE PACKAGE pkg AS
PROCEDURE proc1 (p1 customers.custname%TYPE);
END pkg;
CREATE PACKAGE BODY pkg AS
-- 参数名 p2 与声明的参数名 p1 不一致
PROCEDURE proc1 (p2 customers.custname%TYPE) IS
BEGIN
RAISE NOTICE '%',('custname is ' || p1);
END;
END pkg;
```

程序运行结果如下。

```
ERROR: procedure "proc1" is declared in package "pkg"
and must be defined in the package body
```

正确示例的程序代码如下。

```
CREATE PACKAGE BODY pkg AS
PROCEDURE proc1 (p1 customers.custname%TYPE) IS
    BEGIN
        RAISE NOTICE '%',('custname is ' || p1);
    END;
END pkg;
```

程序运行结果如下。

```
CREATE PACKAGE BODY
```

12.2.2 包的创建

包规范用于声明包的公有组件,如变量、常量、自定义数据类型、异常、过程、函数和游标等。包规范中定义的公有组件不仅可以在包内使用,还可以在包外其他过程和函数中使用。

需要注意的是,为了实现信息的隐藏,不要将所有组件都放在包规范处声明,只应把公有组件放在包规范部分。包的名称是唯一的,但两个包中公有组件的名称可以相同,用"包名.公有组件名"加以区分。

包体是包的具体实现细节,实现在包规范中声明的所有公有过程、函数、游标和变量等,也可以在包体中声明仅属于自己的私有过程、函数和游标等。创建包体时,需要注意以下几点。

(1) 包体只能在包规范被创建或编译后才能创建或编译。

(2) 在包体中实现的过程、函数、游标的名称必须与包规范中的过程、函数、游标一致,包括名称、参数的名称以及参数的模式(IN、OUT、IN OUT)。

(3) 在包体中声明的数据类型、变量、常量都是私有的,只能在本包体中使用,而不能被包体外的应用程序访问和使用。

(4) 包体可以有一个初始化部分,用于对包规范、包体中声明的公有或私有变量进行初始化或其他设置。

创建包规范的语法格式如下。

```
CREATE [OR REPLACE] PACKAGE package_name
  [AUTHID {CURRENT_USER | DEFINER}]
  {IS | AS}
  [公有数据类型定义 [公有数据类型定义]…]
  [公有游标声明 [公有游标声明]…]
  [公有变量、常量声明 [公有变量、常量声明]…]
  [公有函数声明 [公有函数声明]…]
  [公有过程声明 [公有过程声明]…]
END [package_name];
```

创建包体的语法格式如下。

```
CREATE [OR REPLACE] PACKAGE BODY package_name
  {IS | AS}
  [私有数据类型定义 [私有数据类型定义]…]
  [私有变量、常量声明 [私有变量、常量声明]…]
  [私有异常错误声明 [私有异常错误声明]…]
  [私有函数声明和定义 [私有函数声明和定义]…]
  [私有函数过程声明和定义 [私有函数过程声明和定义]…]
  [公有游标定义 [公有游标定义]…]
  [公有函数定义 [公有函数定义]…]
  [公有过程定义 [公有过程定义]…]
BEGIN
  执行部分 (初始化部分)
END package_name;
```

示例 12.2:创建包 emp_good。

功能描述:利用 goods 表并创建一个表 goods_log 和一个包 emp_good,然后从一个匿名块中调用包中的子程序,包 emp_good 既有包规范又有包体。

包 emp_good 的包规范声明了一个公共类型、游标和异常,以及 3 个公共子程序。其中一个公共子程序被重载。

包体定义了包规范中声明的子程序，声明并定义了一个私有变量、一个私有函数，并拥有一个初始化块。

初始化部分（仅在匿名块第一次引用包时运行）向表 goods_log 中插入一行数据。

此程序包可以实现向 goods 表中插入货物、价格更新以及删除货物操作。

程序代码如下。

```
-- 创建表：
DROP TABLE IF EXISTS goods_log;
CREATE TABLE goods_log (
  date_of_action  DATE,
  goodid      VARCHAR2(20),
  package_name  VARCHAR2(30)
);

-- 包规范
CREATE OR REPLACE PACKAGE emp_good AUTHID DEFINER AS
  --声明公共类型、游标、异常：
  TYPE RecTyp IS record (
  goodid   goods.goodid%TYPE,
  price   goods.price%TYPE );

  CURSOR desc_price RETURN RecTyp IS
    SELECT goodid, price
    FROM goods
    ORDER BY price DESC;
  invalid_price EXCEPTION;

  --声明公共子程序：
  FUNCTION hire_good (
    goodid   goods.goodid%TYPE,
    goodname  goods.goodname%TYPE,
    price   goods.price%TYPE
  ) RETURN NUMBER;

  -- 重载的公共子程序：
  PROCEDURE fire_good (emp_goodid NUMBER);
  PROCEDURE fire_good (emp_goodname VARCHAR2);
  PROCEDURE raise_price (emp_goodid NUMBER, emp_price NUMBER);

  END emp_good;
-- 包体：
CREATE OR REPLACE PACKAGE BODY emp_good AS
  number_hired  NUMBER;   --私有变量,仅包体可见
  --定义在包中声明的子程序：
  FUNCTION hire_good (
    goodid    goods.goodid%TYPE,
    goodname  goods.goodname%TYPE,
```

```
    price       goods.price%TYPE
) RETURN NUMBER
IS
BEGIN
  INSERT INTO goods (
    goodid,
    goodname,
    price
  )
  VALUES (
    hire_good.goodid,
    hire_good.goodname,
    hire_good.price
  );
  raise notice '%', ('The number of good is '
        || TO_CHAR(hire_good.goodid));
  RETURN hire_good.goodid;
END hire_good;

PROCEDURE fire_good (emp_goodid NUMBER) IS
BEGIN
  DELETE FROM goods WHERE goodid = emp_goodid;
END fire_good;

PROCEDURE fire_good (emp_goodname VARCHAR2) IS
BEGIN
  DELETE FROM goods WHERE goodname = emp_goodname;
END fire_good;

--私有函数,仅包体可见:
FUNCTION price_ok (
  price NUMBER
) RETURN BOOLEAN
IS
BEGIN
  RETURN (price >= 0);
END price_ok;

PROCEDURE raise_price (
  emp_goodid NUMBER,
  emp_price NUMBER
)
IS
BEGIN
  IF price_ok(emp_price) THEN      --调用私有函数
    UPDATE goods
    SET price = emp_price
    WHERE goodid = emp_goodid;
  ELSE
    RAISE invalid_price;
```

```
      END IF;
    EXCEPTION
      WHEN invalid_price THEN
        raise notice '%', ('The price is out of line with the specification.');
    END raise_price;

BEGIN          --包体的初始化块
    INSERT INTO goods_log (date_of_action, goodid, package_name)
    VALUES (SYSDATE, USER, 'EMP_GOOD');
END emp_good;

-- 在匿名块中调用包中子程序:
DECLARE
  new_emp_goodid NUMBER(6);
BEGIN
  new_emp_goodid := emp_good.hire_good (
     1234, '鼠标', 600
  );
  raise notice '%', ('The goodid is ' || TO_CHAR(new_emp_goodid));
  emp_good.raise_price (new_emp_goodid, 200);
  emp_good.fire_good(new_emp_goodid);
-- 也可以调用以下子程序:
  emp_good.fire_good('computer');
END;
```

程序运行结果如下。

```
The number of good is 1234
The goodid is 1234
```

12.3 程序包的使用

12.3.1 包元素的调用规则

包中可以包含变量、常量、游标、存储过程、函数和异常等对象,其中变量、常量和游标在 SQL 中无法引用,只能用于 PL/SQL 块中。包的初始化块也不能被直接调用,只是在第一次引用包内对象时被执行一次,且仅一次。

(1) 在本包中引用自身的变量和函数。

在初始化代码段、包函数或包存储过程中引用本包中变量的格式如下。

```
[ package_name .] variable
```

在初始化代码段、包函数或包存储过程中引用本包中存储过程或函数的格式如下。

```
[ package_name .] func_name()
[ package_name .] proc_name()
```

（2）在普通的存储过程、函数和其他包中引用包中对象。

在普通的存储过程、函数和其他包中引用包中变量的格式如下。

```
package_name.variable
```

在普通的存储过程、函数和其他包中引用包中存储过程、函数的格式如下。

```
package_name.func_name()
package_name.proc_name()
```

在 SQL 中调用包存储过程或包函数同普通的存储过程或函数的调用相同，但是需要用包名来限定，如下代码。

```
SELECT package_name.func_name();
CALL package_name.proc_name();
```

综上所述，引用包里的元素时需要遵守以下规则。

（1）在包规范中声明的任何元素都是公有的，在包外部可见。

（2）当要在包的外面（一个外部程序中）引用在包规范中定义的元素时，必须使用 package_name.element_name 这种句点表示法格式。

（3）如果要在包的内部（包规范或包体中）使用一个包元素，则不需要包含包的名字，可以直接通过元素名调用，PL/SQL 自动按照包的作用范围解析引用。但是，在包体中定义却没有在包规范中声明的元素则是私有的，只能在包体内部引用。

12.3.2 包数据

包数据是由定义在包级别的变量和常量组成的，也就是说，不是在包的某个函数或过程中定义的。因此包数据的作用范围不仅限于包里的单个程序，而是整个包。在 PL/SQL 运行架构中，包数据结构在整个会话生命周期中是持久性的。

如果一个包数据是在包规范中声明的，则这个数据会在整个会话生命周期内具有持久性，并且可以被那些对这个包有 EXECUTE 权限的程序所使用（公有数据部分）。

如果一个包数据是在包体里声明的，则这个数据也会在整个会话生命周期内具有持久性，但只能被定义在包内部的成员使用（私有数据部分）。

程序包类似 PL/SQL 环境中的全局变量。一个包可以有两类数据结构（公有和私有），对应的也有两种类型的全局包数据：全局公有数据和全局私有数据。这里主要介绍全局公有数据。

在包规范中声明的所有数据结构都属于全局公有的数据结构，也就是位于包以外的程序都可以使用它们。只要在声明语句中没有被声明成常量（CONSTANTS），就可以修改这些全局公有的数据结构。

12.3.3 包游标

显式游标是一种包数据，可以在程序包中声明一个游标——既可以在包体中声明，也可以在包规范中声明。这个游标的状态（包括打开还是关闭，结果集的位置指针）会在整个会话中保持，就像其他的包数据一样。这意味着可以在第一个程序中打开包游标，在第二个程序中提取数据，最后在第三个程序中关闭这个游标。

1．声明包游标

在包规范中声明一个显示游标，与在一个局部 PL/SQL 块中声明游标完全一样。

示例 12.3：在包规范中声明包游标。

功能描述：在包规范中声明一个带有查询语句的完整游标。

程序代码如下。

```
CREATE OR REPLACE PACKAGE good_info
IS
--这是一个非常典型的显式游标定义,在包规范中进行了完整的定义
    CURSOR bygoodname_cur (
        goodname_filter_in IN goods.goodname%TYPE
    )
    IS
      SELECT *
      FROM goods
      WHERE goodname LIKE upper (goodname_filter_in);
    END good_info;
```

接续示例为包体的实现。

示例 12.4：包体实现。

程序代码如下。

```
CREATE OR REPLACE PACKAGE BODY good_info
IS
CURSOR bygoodname_cur (
    goodname_filter_in IN goods.goodname%TYPE
) RETURN goods%ROWTYPE
IS
    SELECT *
    FROM goods
    WHERE goodname LIKE upper (goodname_filter_in);
END good_info;
```

2．包游标调用

利用包游标只需要在游标的名字前面加上包名。若想得到所有和小松鼠有关的货物信息，示例如下。

示例 12.5：查找货物名称中带有小松鼠的货物信息。

功能描述：根据模糊查询％　％查找货物名称中带有"小松鼠"的货物信息。

程序代码如下。

```
DECLARE
  onegood good_info.bygoodname_cur%ROWTYPE;
BEGIN
```

```
  OPEN good_info.bygoodname_cur ('%小松鼠%');

  LOOP
    FETCH good_info.bygoodname_cur INTO onegood;
    EXIT WHEN good_info.bygoodname_cur%NOTFOUND;
    RAISE NOTICE '%',onegood.goodname;
  END LOOP;

  CLOSE good_info.bygoodname_cur;
END;
```

有关包游标的持久性,需要遵循以下规则。

(1) 永远不要假设一个包游标已经关闭了(即已经为再次打开做好准备)。

(2) 永远不要假设一个包游标是打开的(即已经为再次关闭做好准备)。

(3) 确保每次使用完包游标后,总是显式地关闭它。即使在异常处理部分也要加上这个逻辑,这样可以保证无论从哪个点退出程序,游标都会关闭。

12.3.4 查看程序包信息

创建包与创建函数类似,都需要将相关的信息持久化,需要向系统表中插入一些必要的信息,这些信息包括包对象的信息、包中子对象的信息和依赖关系。可以通过下面的语句进行查询,程序代码如下。

```
SELECT * FROM sys_package;
```

程序运行结果如下。

oid	pkgname	pkgnamespace	pkgowner	pkgsecdef	…
13293	standard	11	10	false	…
13394	dbms_sql	11	10	false	…
13637	dbms_output	8000	10	false	…
13657	dbms_ddl	8000	10	false	…
13665	dbms_utility	8000	10	false	…
13668	owa_util	8000	10	false	…
17626	dbms_metadata	16532	10	false	…
17633	pkg	16532	10	false	…
17637	emp_good	16532	10	false	…
17669	good_prices	16532	10	false	…
17782	good_info	16532	10	false	…

引用包变量和类型编译时,需要对相关对象的合法性进行检查。可以使用下面的语句进行查询。

```
SELECT * FROM sys_pkgitem;
```

程序运行结果如下。

```
oid   | pvname          | pkgoid | pvtypoid | ... | pvnotnull | ...
------+-----------------+--------+----------+-----+-----------+----
13294 | dup_val_on_index| 13293  | 0        |     | false     |...
13395 | no_data_found   | 13293  | 0        |     | false     |...
13396 | invalid_cursor  | 13293  | 0        |     | false     |...
13397 | too_many_rows   | 13293  | 0        |     | false     |...
...
```

当要查看程序包的相关信息时，可以使用 DBMS_METADATA 获取包的信息。

示例 12.6：获取包的信息。

功能描述：引入 dbms_metadata 插件，以查询包 emp_good 为例，执行语句，获得包信息。

程序代码如下。

```
CREATE EXTENSION dbms_metadata;
--查找 package 头信息
select dbms_metadata.GET_DDL('package', 'emp_good');
--查找 package 体信息
select dbms_metadata.GET_DDL('package_body', 'emp_good');
--查找 package 头+体信息
select dbms_metadata.GET_DDL('package_all', 'emp_good');
--指定模式
select dbms_metadata.GET_DDL('package_all', 'emp_good', 'sales');
```

程序运行结果如下。

```
    "get_ddl"                                                        |
---------------------------------------------------------------------+
CREATE OR REPLACE PACKAGE EMP_GOOD AS
  CURSOR DESC_PRICE RETURN RECTYP IS
SELECT GOODID, PRICE
FROM GOODS
    ORDER BY PRICE DESC;
INVALID_PRICE EXCEPTION;
FUNCTION HIRE_GOOD (
    GOODID   GOODS.GOODID%TYPE,
    GOODNAME  GOODS.GOODNAME%TYPE,
    PRICE    GOODS.PRICE%TYPE
 ) RETURN NUMBER;
PROCEDURE FIRE_GOOD (EMP_GOODID NUMBER);
PROCEDURE FIRE_GOOD (EMP_GOODNAME VARCHAR2);
PROCEDURE RAISE_PRICE (EMP_GOODID NUMBER, EMP_PRICE NUMBER);
END;
CREATE OR REPLACE PACKAGE BODY SALES.EMP_GOOD AS

NUMBER_HIRED  NUMBER;      --私有变量,仅包体可见
-- 定义在包中声明的子程序:
  FUNCTION HIRE_GOOD (
```

```
    GOODID      GOODS.GOODID%TYPE,
    GOODNAME    GOODS.GOODNAME%TYPE,
    PRICE       GOODS.PRICE%TYPE
  ) RETURN NUMBER
  IS
  BEGIN
  INSERT INTO GOODS (
      GOODID,
      GOODNAME,
      PRICE
  )
VALUES (
HIRE_GOOD.GOODID,
HIRE_GOOD.GOODNAME,
HIRE_GOOD.PRICE
  );
RAISE NOTICE '%', ('THE NUMBER OF GOOD IS '
          || TO_CHAR(HIRE_GOOD.GOODID));
RETURN  HIRE_GOOD.GOODID;
END HIRE_GOOD;

PROCEDURE FIRE_GOOD (EMP_GOODID NUMBER) IS
BEGIN
DELETE FROM GOODS WHERE GOODID = EMP_GOODID;
END FIRE_GOOD;
PROCEDURE FIRE_GOOD (EMP_GOODNAME VARCHAR2) IS
BEGIN
    DELETE FROM GOODS WHERE GOODNAME = EMP_GOODNAME;
END FIRE_GOOD;
-- 私有函数,仅包体可见:
FUNCTION PRICE_OK (
PRICE NUMBER
) RETURN BOOLEAN
IS
BEGIN
RETURN (PRICE >= 0);
END PRICE_OK;
PROCEDURE RAISE_PRICE (
  EMP_GOODID NUMBER,
  EMP_PRICE NUMBER
  )
  IS
  BEGIN
IF PRICE_OK(EMP_PRICE) THEN       --调用私有函数
     UPDATE GOODS
  SET PRICE = EMP_PRICE
  WHERE GOODID = EMP_GOODID;
    ELSE
     RAISE INVALID_PRICE;
 END IF;
```

```
EXCEPTION
    WHEN INVALID_PRICE THEN
      RAISE NOTICE '%', ('THE PRICE IS OUT OF LINE WITH THE SPECIFICATION.');
END RAISE_PRICE;
BEGIN      --包体的初始化块
    INSERT INTO GOODS_LOG (DATE_OF_ACTION, GOODID, PACKAGE_NAME)
    VALUES (SYSDATE, USER, 'EMP_GOOD');
END EMP_GOOD ; "
```

要查找指定包中所有函数的信息(函数名、函数类别、函数参数信息、函数体等),可以使用 sys_get_package_function_detail 语句。

示例 12.7：查找指定包中所有函数的信息。

功能描述：执行 select * from sys_get_package_function_detail('参数1','参数2')。

参数1：包所在的模式名,模式名为 NULL 时,指定为当前模式。

参数2：要查找的包名。

程序代码如下。

```
--不指定模式
select proname,
    prokind,
    pronargs,
    proargnames,
    proargmodes,
    proargtypes,
    prorettype,
    prosrc
    from sys_get_package_function_detail(null, 'emp_good');
--指定模式
select proname,
    prokind,
    pronargs,
    proargnames,
    proargmodes,
    proargtypes,
    prorettype,
    prosrc
    from sys_get_package_function_detail('sales', 'emp_good');
```

程序运行结果(不指定模式下查询包的函数信息)如下。

```
 proname    | prokind | pronargs | proargnames              |…| prosrc                     |
------------+---------+----------+--------------------------+--+----------------------------+
 hire_good  | f       | 3        | {goodid,goodname,price}  |…| FUNCTION hire_good …       |
 fire_good  | p       | 1        | {emp_goodid}             |…| PROCEDURE fire_good…       |
 fire_good  | p       | 1        | {emp_goodname}           |…| PROCEDURE fire_good …      |
 raise_price| p       | 2        | {emp_goodid,emp_price}   |…| PROCEDURE raise_price…     |
```

程序运行结果(指定模式下查询包的函数信息)如下。

```
 proname    | prokind | pronargs | proargnames            | … | prosrc
------------+---------+----------+------------------------+---+------------------------
 hire_good  | f       | 3        | {goodid,goodname,price}| … | FUNCTION hire_good …
 fire_good  | p       | 1        | {emp_goodid}           | … | PROCEDURE fire_good …
 fire_good  | p       | 1        | {emp_goodname}         | … | PROCEDURE fire_good …
 raise_price| p       | 2        | {emp_goodid,emp_price} | … | PROCEDURE raise_price …
```

第 13 章

触 发 器

触发器是数据库提供给用户,用以保证数据完整性的一种方法。它是和事件相关的特殊的过程或函数,只要发生指定事件,数据库就会自动调用它。本章包括以下主要内容。
- 触发器简介。
- DML 触发器。
- 事件触发器。
- 触发器设计注意事项。
- 触发器管理。

13.1 触发器简介

13.1.1 触发器的概念

与存储过程相同的是,触发器是一个命名的 PL/SQL 单元。它存储在数据库中,并且可以重复调用。与存储过程不同的是,触发器存在启用和禁用两种状态,并且不能被显式地调用。

触发器被启用时,只要它的触发事件发生,触发器就会被触发,即数据库会自动调用它。当触发器被禁用时,不会被触发。

使用 CREATE TRIGGER 语句创建触发器。可以根据触发语句及其作用的数据库对象来指定触发事件。触发器是在数据库对象上创建或定义的,数据库对象可以是表、视图、模式或数据库。还可以指定时间点,它确定触发器是在触发语句运行之前还是之后触发,以及是否针对触发语句影响的每一行触发。触发器创建完成后,默认是启用状态。

需要注意的是,虽然触发器和约束都可以约束数据输入,但它们有以下很大的不同。

(1)触发器仅适用于新数据。例如,触发器可以阻止 DML 语句将 NULL 值插入数据库列,但该列可能包含 NULL 在定义触发器之前或禁用触发器时插入到列中的值。

(2)约束可以应用于新数据或同时应用于新数据和现有数据。约束行为取决于约束状态。

(3)与强制执行相同规则的触发器相比,约束更容易编写,且不易出错。但是,触发器可以强制执行一些约束不能执行的复杂业务规则。当强制执行无法使用约束定义的复杂业务时,建议使用触发器,用以约束数据输入。

13.1.2 触发器的作用

可以使用触发器来自定义数据库管理系统。例如，可以使用触发器实现以下任务。

(1) 记录事件。

(2) 收集有关表访问的统计信息。

(3) 在针对视图发出 DML 语句时修改表数。

(4) 将有关数据库事件、用户事件和 SQL 语句的信息发布到订阅应用程序。

(5) 防止在正常工作时间后对表进行 DML 操作。

(6) 强制执行无法使用约束定义的复杂业务或参照完整性规则。

13.1.3 触发器的种类

根据触发器作用的对象以及触发事件的不同，可以分为以下两大类。

(1) DML 触发器：建立在表或视图上，触发事件由 INSERT、UPDATE、DELETE 等 DML 语句组成。

(2) 事件触发器：建立在模式或数据库上，触发事件由 DDL（CREATE、ALTER、DROP）组成。

13.2 DML 触发器

13.2.1 DML 触发器的用途

DML 触发器是在表或视图上创建的基于 DML 操作所建立的触发器。其常见用途包括以下内容。

(1) 记录并审核对表中数据的修改操作，实现审计功能。

(2) 实现比 CHECK 约束更加复杂的完整性约束，如禁止非业务时间的数据操作。

(3) 实现某种业务逻辑，如在增加或删除员工时自动更新部门中的人数。

(4) 使用触发器生成序列号的值，为字段提供默认的数据。

(5) 实现数据的同步复制。

虽然触发器可以实现很多功能，但是它也存在以下一些缺点。

(1) 触发器增加了数据库结构的复杂度和系统维护的难度。

(2) 触发器由数据库服务器运行，需要占用更多的数据库资源。

(3) 触发器不能接收参数，只能基于当前的操作对象触发。

13.2.2 创建 DML 触发器

触发器由触发器头部和触发器体两部分组成，触发器头部包括以下内容。

(1) 作用对象：触发器作用的对象包括表、视图。

(2) 触发事件：激发触发器执行的事件，如 DML、DDL、数据库事件等，可以将多个事件用关系运算符 OR 组合。

(3) 触发时间：用于指定触发器在触发语句完成前或完成后执行。若指定为 AFTER，则表示先执行触发语句,然后执行触发器;若指定为 BEFORE,则表示先执行触发器,然后执行触发语句。

(4) 触发级别:用于指定触发器响应触发事件的方式。默认为语句级触发器,即触发事件发生后,触发器只执行一次。如果指定为 FOR EACH ROW,即为行级触发器,则触发事件每作用于一条记录,触发器就执行一次。

(5) 触发条件:由 WHEN 子句指定一个逻辑表达式,当触发事件发生,而且 WHEN 条件为 TRUE 时,触发器才被执行。

触发器体指触发器执行时进行的操作,即触发器的行为,可以是 PL/SQL 块、触发器函数、存储过程和函数。

如果是在不可更新的视图上创建的触发器,则称为 INSTEAD OF 触发器。对视图的 DML 操作,最终都会转变为对基本表的操作。

语法格式如下。

```
create [or replace] trigger trigger_name
{before|after}
{delete|insert|update|[of 列名]}
on 表名
[for each row [when(条件)]]
[
  --触发器逻辑--
];
```

DML 触发器的触发事件可能由多个触发语句组成。当触发其中一个触发器时,可以通过表 13.1 的条件谓词确定使用哪一个触发器。

表 13.1 条件谓词及其适用条件

条 件 谓 词	条　　件
INSERTING	一个 INSERT 语句触发了触发器
UPDATING	一个 UPDATE 语句触发了触发器
UPDATING('column')	影响指定列的 UPDATE 语句触发了触发器
DELETING	一个 DELETE 语句触发了触发器

条件谓词可以出现在 BOOLEAN 表达式出现的任何地方。

示例 13.1:创建 DML 触发器。

功能描述:此示例创建 1 个 DML 触发器,该触发器使用条件谓词确定 4 个可能的触发语句中哪 1 个触发了它。

程序代码如下。

```
CREATE OR REPLACE TRIGGER tri
  BEFORE
    INSERT OR
    UPDATE OF addrname, parentid OR
    DELETE
```

```
      ON adminaddrs
BEGIN
  CASE
    WHEN INSERTING THEN
      RAISE NOTICE 'Inserting';
    WHEN UPDATING('addrname') THEN
      RAISE NOTICE 'Updating addrname';
    WHEN UPDATING('parentid') THEN
      RAISE NOTICE 'Updating parentid ID';
    WHEN DELETING THEN
      RAISE NOTICE 'Deleting';
  END CASE;
END;
```

示例 13.2：触发器派生新列值。

功能描述：每当插入或更新行时，本例中的触发器都会为表派生新的列值。需要对 customers 表结构进行如下更改。

程序代码如下。

```
ALTER TABLE customers ADD(Uppercustname VARCHAR2(20));

CREATE OR REPLACE TRIGGER Derived
  BEFORE INSERT OR UPDATE OF custname ON customers
FOR EACH ROW
BEGIN
  :NEW.Uppercustname := UPPER(:NEW.custname);
END;
```

13.2.3 触发器体

KingbaseES 中触发器的行为，除了可以是 PL/SQL 块，还可以是触发器函数，专门用于触发器响应后执行，语法格式如下。

```
CREATE OR REPLACE FUNCTION function_name()
RETURNS TRIGGER        --返回一个触发器
AS
BEGIN
  ...
END;
```

一个触发器函数被调用时，顶层块会自动创建一些特殊变量，供其使用，如表 13.2 所示。

表 13.2 触发器函数中的变量及其说明

变量名称	数据类型	说　　明
NEW	RECORD	行级触发器中的 INSERT/UPDATE 操作保持新数据行
OLD	RECORD	行级触发器中的 UPDATE/DELETE 操作保持新数据行

续表

变量名称	数据类型	说明
TG_NAME	name	实际触发的触发器名
TG_WHEN	text	值为 BEFORE、AFTER 或 INSTEAD OF 的一个字符串,取决于触发器的定义
TG_LEVEL	text	值为 ROW 或 STATEMENT 的一个字符串,取决于触发器的定义
TG_OP	text	值为 INSERT、UPDATE、DELETE 或 TRUNCATE 的一个字符串,说明触发器为哪个操作引发
TG_RELID	oid	导致触发器调用的表的对象 ID
TG_TABLE_NAME	name	导致触发器调用的表的名称
TG_TABLE_SCHEMA	name	导致触发器调用的表所在的模式名
TG_NARGS	integer	在 CREATE TRIGGER 语句中给触发器函数的参数数量
TG_ARGV[]	text 数组	来自 CREATE TRIGGER 语句的参数。索引从 0 开始计数。非法索引(小于 0 或大于等于 tg_nargs)会导致返回一个空值

其中,OLD、NEW 也称为伪记录,因为它们具有记录结构,但允许在比记录更少的上下文中使用。伪记录的结构是 table_name%ROWTYPE,其中 table_name 是创建触发器的表名称。

在简单触发器的 trigger_body 中,相关名称是绑定变量的占位符。使用以下语法引用伪记录的字段。

```
:pseudorecord_name.field_name
```

默认地,OLD 代表老数据,NEW 代表新数据,表 13.3 展示了二者在不同 DML 语句中的区别。

表 13.3 新老数据在不同 DML 语句中的区别

触发语句	OLD.field 值	NEW.field 值
INSERT	NULL	插入后值
UPDATE	更新前值	更新后值
DELETE	预删除值	NULL

示例 13.3:触发器体为 PL/SQL 块,记录对 supply.totlwhamt 的更改。

功能描述:本例创建 1 个日志表和 1 个 UPDATE 触发器,在任何语句影响 supply 表的 totlwhamt 列后,该触发器在日志表中将插入 1 行,然后更新 supply.totlwhamt,并显示日志表。

程序代码如下。

```
CREATE TABLE supply_log (
   shopid    bpchar(6),
```

```
    goodid      bpchar(12),
    log_date    DATE,
    new_totlwhamt INT,
    action      varchar2(20));
```

supply.totlwhamt 更新后,在日志表中创建插入行触发器,程序代码如下。

```
CREATE OR REPLACE TRIGGER log_supply_increase
  AFTER UPDATE OF totlwhamt ON supply
  FOR EACH ROW
BEGIN
  INSERT INTO supply_log (shopid, goodid, log_date, new_totlwhamt, action)
  VALUES (:NEW.shopid,:NEW.goodid, SYSDATE, :NEW.totlwhamt, 'new totlwhamt');
END;
```

更新 supply.totlwhamt,程序代码如下。

```
UPDATE supply
SET totlwhamt=totlwhamt+10
WHERE totlwhamt=(SELECT min(totlwhamt) FROM supply)
```

显示日志表,程序代码如下。

```
SELECT * FROM supply_log;
```

程序运行结果如下。

```
 shopid | goodid    | log_date            | new_totlwhamt | action        |
--------+-----------+---------------------+---------------+---------------+
 492955 |607680932  | 2022-09-22 15:36:37 |10.00          | new totlwhamt |
```

示例 13.4:触发器体中调用触发器函数,打印库存变化信息。

功能描述:本例在触发器体中调用触发器函数,只要 DELETE、INSERT 或 UPDATE 语句影响 supply 表,该触发器就会打印库存变化信息。

程序代码如下。

```
CREATE OR REPLACE FUNCTION process_supply_audit() RETURNS TRIGGER
AS
DECLARE
  totlwhamt_diff NUMERIC(5,2);
BEGIN
  IF (TG_OP = 'DELETE') THEN
     RAISE NOTICE '%,shopid=%,goodid=%',TG_OP,OLD.shopid,OLD.goodid;
  ELSIF (TG_OP = 'UPDATE') THEN
    totlwhamt_diff:=:NEW.totlwhamt-:OLD.totlwhamt;
    RAISE NOTICE '%,shopid=%,goodid=%,old.totlwhamt=%,new.totlwhamt=%',
      TG_OP,NEW.shopid,NEW.goodid,OLD.totlwhamt,NEW.totlwhamt;
    RAISE NOTICE 'Difference: %' ,totlwhamt_diff;
```

```
    ELSIF (TG_OP = 'INSERT') THEN
      RAISE NOTICE '%,shopid=%,goodid=%',TG_OP,NEW.shopid,NEW.goodid;
    END IF;
    RAISE NOTICE '%',now();
    RETURN NULL;
END;

CREATE OR REPLACE TRIGGER supply_audit
AFTER INSERT OR UPDATE OR DELETE ON supply
FOR EACH ROW EXECUTE FUNCTION process_supply_audit();
```

触发语句,程序代码如下。

```
UPDATE supply
SET totlwhamt=totlwhamt-1
WHERE shopid='219442' AND
      goodid='7532692'
```

程序运行结果如下。

```
UPDATE,shopid=219442,goodid=7532692 ,old.totlwhamt=9993,new.totlwhamt=9992
Difference: -1.00
2023-02-02 18:38:31.342898+08
```

13.2.4 INSTEAD OF 触发器

1. INSTEAD OF 触发器概述

INSTEAD OF 触发器是在不可更新的视图上创建的触发器,响应 DML 操作。对视图的 DML 操作最终都会转变为对基本表的操作。所以这种类型的触发器被称为 INSTEAD OF 触发器。

INSTEAD OF DML 触发器的主要作用是修改一个本来不可更新的视图,不可更新的视图如果包含了下列任意一项,则视图就不可以更改。

（1）涉及多个表的连接操作。

（2）DISTINCT 操作符。

（3）GROUP BY 子句。

（4）聚集函数（AVG、SUM 等）。

（5）集合操作符（UNION、UNION ALL 等）。

2. 创建 INSTEAD OF 触发器

创建（或者修改）一个 INSTEAD OF 触发器的语法格式如下。

```
CREATE [OR REPLACE] TRIGGER trigger_name
INSTEAD OF operation
ON view_name
FOR EACH ROW
PL/SQL 块
```

示例 13.5：INSTEAD OF 触发器。

功能描述：本例创建视图 shopstore_info，以显示有关店铺的信息。视图在本质上不可更新，该示例创建 1 个 INSTEAD OF 触发器，用以处理 INSERT 指向视图的语句。触发器将行插入到视图的基本表 shopstores 中。

程序代码如下。

```
CREATE OR REPLACE VIEW shopstore_info AS
    SELECT  shopid, shopname, shopurl, custgrading, delygrading , servgrading
,comprgrading
    FROM shopstores
    WHERE shopname ='KingbaseES 旗舰店';

CREATE OR REPLACE TRIGGER shopstore_info_insert
    INSTEAD OF INSERT ON shopstore_info FOR EACH ROW
    BEGIN
      INSERT INTO shopstores(shopid,shopname,shopurl,custgrading,
               delygrading ,servgrading ,comprgrading)
      VALUES (
        :new.shopid,
        :new.shopname,
        :new.shopurl,
          :new.custgrading,
          :new.delygrading,
          :new.servgrading,
          :new.comprgrading);
    END shopstore_info_insert;
```

13.2.5 触发器触发的顺序

如果为同一张表上的同一语句定义了两个或多个具有不同时间点的触发器，则它们将按以下顺序触发。

（1）所有 BEFORE STATEMENT 触发器。
（2）所有 BEFORE EACH ROW 触发器。
（3）所有 AFTER EACH ROW 触发器。
（4）所有 AFTER STATEMENT 触发器。

13.3 事件触发器

13.3.1 事件触发器概述

事件触发器是执行 DDL 语句时被触发的触发器，DDL 语句是用来创建或修改数据库对象的语句，如 CREATE TABLE、ALTER INDEX、DROP TRIGGER 等。

事件触发器可以用于监控和防止针对数据库对象的修改命令。例如，为了防止恶意攻击或者用户误操作导致删除数据表，就可以在系统日常运行中禁止执行某些数据库命令，如 DROP TABLE 语句。

13.3.2 创建事件触发器

创建事件触发器的语法和创建 DML 触发器的语法类似,只是触发事件不同,而且事件触发器不是作用在某张表上,语法格式如下。

```
CREATE EVENT TRIGGER
ON event
[ WHEN filter_variable IN (filter_value [,...]) [ and ...]]
EXECUTE PROCEDURE function_name()
```

创建事件触发器之前,必须先创建事件触发器函数,该函数的返回类型为 event_trigger。

示例 13.6:事件触发器禁止所有 DDL 语句。

程序代码如下。

```
CREATE OR REPLACE FUNCTION abort_any_command()
returns event_trigger
language plpgsql
AS $$
BEGIN
    RAISE EXCEPTION 'command % is disabled', tg_tag;
END;
$$;

CREATE EVENT TRIGGER abort_DDL ON DDL_command_start
EXECUTE PROCEDURE abort_any_command();
```

执行 DDL 语句将会报错,程序代码如下。

```
DROP TABLE adminaddrs;
```

程序运行结果如下。

```
ERROR:command DROP TABLE is disabled
```

如果希望再次允许 DDL 操作,可以禁止该触发器,程序代码如下。

```
ALTER EVENT TRIGGER abort_ddl DISABLE;
```

13.4 触发器设计注意事项

定义和设计触发器时,需要注意以下事项。

(1)触发器的使用需要确保无论何时发生指定的事件,都会执行必要的操作(无论哪个用户或应用程序发出触发语句)。例如,使用触发器确保每当有用户修改表时,都会更新日志文件。

（2）不要使用触发器重复实现数据库的某些特性。例如，如果可以使用约束执行相同操作，就不要创建触发器禁止无效数据。

（3）不要创建依赖于 SQL 语句处理行顺序（该顺序是可以变化的）的触发器。例如，如果变量的当前值取决于行触发器正在处理的行，则不要将值分配给行触发器中的全局包变量。如果触发器更新全局包变量，需要在 BEFORE 语句触发器中初始化这些变量。

（4）BEFORE 在将行数据写入磁盘之前，使用行触发器修改行。

（5）使用 AFTER 行触发器获取行 ID，并在操作中使用它。

（6）如果 BEFORE 语句触发器的触发语句（UPDATE 或 DELETE 语句）与正在运行的 UPDATE 语句冲突，则数据库对 SAVEPOINT 执行透明 ROLLBACK，并重新启动触发语句。在触发语句成功完成之前，数据库可以多次执行此操作。每次数据库重新启动触发语句时，触发器都会触发。ROLLBACK 到 SAVEPOINT 不会撤销对触发器引用的包变量的更改。若要检测这种情况，需要在包中定义一个计数器变量。

（7）不要创建递归调用的触发器，例如，不要创建在定义触发器的表上执行 UPDATE 语句的 AFTER UPDATE 触发器。触发器会递归触发，直到内存不足。

（8）如果创建的触发器包含访问远程数据库的语句，则将该语句的异常处理程序放在存储的子程序中，并从触发器调用子程序。

（9）只有事务提交时，触发器才会触发。

（10）如果允许模块化安装在相同表上且具有触发器的应用程序，则需要创建多个相同类型的触发器，而不是运行一系列操作的单个触发器。每个触发器都会看到先前触发的触发器所作的更改。每个触发器都可以看到 OLD 和 NEW 的值。

13.5 触发器管理

13.5.1 禁用与启用触发器

默认情况下，CREATE TRIGGER 语句创建一个处于启用状态的触发器。如果在禁用状态下创建触发器，则需要指定 DISABLE。在禁用状态下创建触发器可以确保其在启用之前编译无错误发生。

暂时禁用触发器的情况包括以下内容。

（1）触发器引用了一个不可用的对象。

（2）必须进行大量数据加载，并且希望它在不触发触发器的情况下快速进行。

（3）重新加载数据。

启用或禁用单个触发器，语法格式如下。

```
ALTER TRIGGER [schema.]trigger_name { ENABLE | DISABLE };
```

需要启用或禁用在特定表上创建的所有版本的所有触发器，语法格式如下。

```
ALTER TABLE table_name { ENABLE | DISABLE } ALL TRIGGERS;
```

上述两个语句中，schema 是包含触发器的模式的名称，默认是当前模式。

13.5.2 修改、重编译与删除触发器

（1）修改触发器：可以使用 CREATE OR REPLACE TRIGGER 语句完成，也可以先执行 DROP TRIGGER 语句删除触发器，然后使用 CREATE TRIGGER 语句重建触发器。

（2）重编译触发器：如果触发器中引用的其他数据库对象（如子程序、包等）被修改或删除，触发器将失效。只有当下一次触发事件发生时，系统重新编译触发器才能使其重新生效。如果系统对触发器的自动重新编译失败，则需要使用 ALTER TRIGGER…COMPILE 语句重新编译触发器，语法格式如下：

```
ALTER TRIGGER trigger_name COMPILE;
```

（3）删除触发器：可以使用 DROP TRIGGER 语句删除不再需要的触发器，语法格式如下。

```
DROP TRIGGER trigger_name;
```

13.5.3 触发器信息查询

*_TRIGGERS 静态数据字典视图显示有关触发器的信息。

示例 13.7：查看有关触发器的信息。

功能描述：本例创建一个触发器，并查询静态数据字典视图 USER_TRIGGERS 两次。首先显示其类型、触发事件和创建它的表的名称，然后显示其主体。

程序代码如下。

```
CREATE OR REPLACE TRIGGER Cust_count
  AFTER DELETE ON customers
DECLARE
  n INTEGER;
BEGIN
  SELECT COUNT(*) INTO n FROM customers;
  RAISE NOTICE 'There are now % customers.', n;
END;
```

查询的代码如下。

```
SELECT Trigger_type, Triggering_event, Table_name
FROM USER_TRIGGERS
WHERE Trigger_name = 'CUST_COUNT';
```

程序运行结果如下。

```
Trigger_type     | triggering_event |table_name |
-----------------+------------------+-----------+
AFTER STATEMENT  | DELETE           | customers |
```

查询的代码如下。

```
SELECT Trigger_body
FROM USER_TRIGGERS
WHERE Trigger_name = 'CUST_COUNT';
```

程序运行结果如下。

```
trigger_body                |
----------------------------+
CREATE OR REPLACE TRIGGER cust_count AFTER DELETE ON sales.customers FOR
EACH STATEMENT
DECLARE
  n INTEGER;
BEGIN
  SELECT COUNT(*) INTO n FROM customers;
  RAISE NOTICE 'There are now % customers.', n;
END
```

第 14 章

PL/SQL 的代码加密

加密 PL/SQL 源代码就是通过混淆隐藏 PL/SQL 内容的过程。包含加密后内容的文件被称为加密文件,加密文件可以被 KSQL 或导入/导出工具移动、备份以及处理,但内容无法通过数据字典视图 *_SOURCE 查看。对 PL/SQL 源代码进行加密,可以在交付应用时隐藏源码和实现细节,也可以防止发布出去的代码被篡改;KingbaseES 数据库系统内置的 PL/SQL 程序包和类型的代码绝大部分经过了加密处理。KingbaseES 提供了两种加密 PL/SQL 源代码的方法,即 PL/SQL Wrapper 实用程序和 DBMS_DDL 子程序。本章包括以下主要内容。

(1) PL/SQL 代码加密概述。
(2) Wrapper。
(3) DBMS_DDL 程序包的使用。

 14.1　PL/SQL 代码加密概述

PL/SQL 加密可以使用 PL/SQL Wrapper 实用程序或 DBMS_DDL 子程序实现。其中,PL/SQL Wrapper 实用程序从命令行运行,可以加密 SQL 脚本文件中任何可加密的 PL/SQL 对象,例如一个 KSQL 安装脚本。DBMS_DDL 子程序可以加密单个动态生成的 PL/SQL 单元,例如 CREATE PROCEDURE 命令。

两种加密方法都可以检查标记错误(如字符串超长),但是不会检查语法或语义错误(如不存在的表或视图)。

对 PL/SQL 源代码加密的目的是防止其他开发人员误用代码,并能够阻止商业竞争对手等查看其中的算法内容。

对 PL/SQL 源代码加密时,需要注意以下几点。

(1) 可加密包体(仅加密包体,而不是包规范)、类型体、函数、存储过程。

(2) 不加密包规范是允许其他开发人员查看使用包或类型所需的信息。加密包体可以防止开发人员看到包的实现。

(3) 不能编辑加密的文件。如果加密的文件需要更改,必须编辑原始文本,然后将其加密。

(4) 分发加密文件之前,需要在文本编辑器中查看并确保所有重要部分都已加密。

(5) 加密工具无法加密触发器的源代码。如果要隐藏触发器的实现细节，可以将具体实现放入一个存储程序，然后加密该程序，最后编写一个调用加密程序的触发器。

14.2 Wrapper

14.2.1 使用 PL/SQL Wrapper 实用程序

PL/SQL Wrapper 实用程序接收一个 SQL 文件作为输入，加密该文件中可加密的 PL/SQL 对象（不会加密匿名块、触发器或非 PL/SQL 代码），然后输出一个对应的加密文件。

PL/SQL Wrapper 实用程序无法连接到 KingbaseES 数据库。要运行 PL/SQL Wrapper 实用程序，在操作系统提示符下输入以下命令（等号周围无空格）。

语法格式如下。

```
wrap iname=input_file [ oname=output_file ]
```

input_file 是包含 SQL 语句的任意组合的现有文件名称。

output_file 是 PL/SQL Wrapper 实用程序创建的文件名称——加密后的文件。

如果 input_file 是一个加密文件，则 output_file 是再次加密的文件。

input_file 的默认文件扩展名是 sql，output_file 的默认名称是 input_file.plb。因此，以下这些命令是等价的。

```
wrap iname=/kesdir/kesfile
wrap iname=/kesdir/kesfile.sql oname=/kesdir/kesfile.plb
```

14.2.2 PL/SQL Wrapper 实用程序的输入与输出文件

输入文件可以包含任何 SQL 语句的组合，大多数语句不会更改。定义子程序、程序包或对象类型的 CREATE 语句被加密后，它们的程序体被替换为一种 PL/SQL 编译器能够理解的加密形式。

以下 CREATE 语句会被加密。

```
CREATE [OR REPLACE] FUNCTION function_name
CREATE [OR REPLACE] PROCEDURE procedure_name
CREATE [OR REPLACE] PACKAGE package_name
CREATE [OR REPLACE] PACKAGE BODY package_name
CREATE [OR REPLACE] TYPE type_name AS OBJECT
CREATE [OR REPLACE] TYPE type_name UNDER type_name
CREATE [OR REPLACE] TYPE BODY type_name
```

语句 CREATE[OR REPLACE] TRIGGER 以及[DECLARE] BEGIN .. END 块不加密。其他所有 SQL 语句不变，直接写入输出文件。

被加密单元中的所有注释行，除了 CREATE OR REPLACE 头部的注释，或者 C-style 的注释（使用/＊ ＊/界定），都会被删除。

输出文件是一个文本文件,可以像脚本一样在 KSQL 中运行并创建 PL/SQL 过程、函数以及程序包。使用以下命令可以查看加密文件。

```
seamart=# \i wrapped_file_name.plb;
```

14.2.3　PL/SQL Wrapper 加密的优点和局限性

1. 优点

(1) 其他开发人员不易误用你的代码,或者商业竞争对手难以查看你的算法。

(2) 源代码不能通过数据字典视图 USER_SOURCE、ALL_SOURCE 或 DBA_SOURCE 进行查看。

(3) KSQL 能够处理加密的源文件。

(4) Import 和 Export 工具支持加密文件。可以备份或移动加密的程序。

2. 局限性

KingbaseES 中,加密文件只会删除单行注释。

14.2.4　示例

示例 14.1:test1.sql 是一个可加密的 PL/SQL 单元(函数 chksupply_totlwhamt)。

注:该文件还包含 PL/SQL 风格注释和一些 SQL 语句。

程序代码如下。

```
/*创建 1 个函数,根据参数指定的货物编码、商店编码、卖出货物数量,查询计算库存减去指定的
货物数量后的剩余库存数量*/
/*以下 PLSQL 风格注释会被删除*/
CREATE OR REPLACE FUNCTION "chksupply_totlwhamt"
(--创建名为 chksupply_totlwhamt 的函数,模式为 sales
  in s_goodid INTEGER, in s_shopid INTEGER, in saleamt INTEGER
  --依次定义了货物编码、商店编码、卖出货物数量 3 个参数,数据类型为整数
)
RETURNs INTEGER as                  --定义了函数返回类型为整数
DECLARE surplus INTEGER;            --声明了名为 surplus 的整数变量
BEGIN
  SELECT totlwhamt-saleamt
  FROM   "supply"
  WHERE shopid = s_shopid AND goodid = s_goodid INTO surplus;
--查询指定商店的指定货物的库存数量,并减去传入的卖出货物数量
RETURN surplus;--返回剩余库存数量
END;
```

要加密该文件,从操作系统提示符中运行以下命令,进行加密。

```
SQL> wrap iname=test1.sql
```

执行成功后,输出的信息如下。

```
Processing test1.sql to test1.plb
```

若想查看加密后的内容,可以在 KSQL 中运行\i test1.plb 查看,代码如下。

```
seamart=# \i test1.plb
```

加密后的 test1.plb 文件内容如下。

```
/*创建1个函数,根据参数指定的货物编码、商店编码、卖出货物数量,查询计算库存减去指定的
货物数量后的剩余库存数量*/
/*以下 PLSQL 风格注释会被删除*/
CREATE OR REPLACE FUNCTION "chksupply_totlwhamt" (
  in s_goodid INTEGER, in s_shopid INTEGER, in saleamt INTEGER
) RETURNs INTEGER as WRAPPED
C810HCQWCGpd1PiSgX500ibAVLzwrBBTfuE18OOLy3dwEYItPSvYn43tmVj4
3srUFt5keky9LmUxCo41QWjSbB5SQW4raG+DxCZTIH5XxYo51GHUYJGb1GPP
7MFT8ujL2SHHKSI3Vl0w+8yL/7/6sQwBDsAHizQWuNObJ6/kDldL0ed27qNG
sFPUZ1U5kBM86tmxx7l2gP5DcSH+sd1ATQAAAAAAAAAAAAAAAAAAAA=
END;
```

示例 14.2:test2.sql 是一个可加密的 PL/SQL 单元(过程 UPDATEsupply_totlwhamt)。
注:该文件还包含 C-style 风格注释和一些 SQL 语句。
程序代码如下。

```
/*创建1个存储过程,根据参数指定的货物编码、商店编码、卖出货物数量,更新该指定商品的库
存数量*/
/*以下 C-style 风格注释不会被删除*/
CREATE OR REPLACE PROCEDURE "UPDATEsupply_totlwhamt" (
     /*创建名为 UPDATEsupply_totlwhamt 的函数,模式为 test*/
  IN s_goodid INTEGER, IN s_shopid INTEGER, IN s_saleamt INTEGER
     /*依次定义了货物编码、商店编码、卖出货物数量3个参数,数据类型为整数*/
) AS
DECLARE surplus INTEGER;       /*声明了名为 surplus 的整数变量*/
BEGIN
surplus:="chksupply_totlwhamt"(s_goodid,s_shopid,saleamt);
     /*调用上面创建的 chksupply_totlwhamt 函数,并将返回值赋值给 surplus 变量*/
  IF surplus>0              /*判断 surplus 大于 0,则执行更新操作*/
  THEN
  UPDATE "supply" SET
  "totlwhamt" = "totlwhamt" -s_saleamt
  WHERE shopid = s_shopid AND goodid = s_goodid;
  END IF;                   /*结束 IF 判断*/
END;
```

查看加密后的内容,程序代码如下。

```
/*创建1个存储过程,根据参数指定的货物编码、商店编码、卖出货物数量,更新该指定商品的库
存数量*/
/*以下 C-style 风格注释不会被删除*/
```

```
CREATE OR REPLACE PROCEDURE "UPDATEsupply_totlwhamt" (
        /*创建名为 UPDATEsupply_totlwhamt 的函数,模式为 test*/
IN s_goodid INTEGER, IN s_shopid INTEGER, IN s_saleamt INTEGER
        /*依次定义了货物编码、商店编码、卖出货物数量 3 个参数,数据类型为整数*/
) AS WRAPPED
C810HCQWCGpd1PiSgX500o6Z1YxShLzuYCccF1z3BSbt6yAt06ZbAZQPmTUT
5/ZbZNsKXic+dF0ckSGOy4nHjXLjxWxWHy69mumXawKm5jNRIz346bvML6gU
MKv4PczIY7hbs5G4L8Fpa5eOU2LnCld2+tBB8noZr9+QlraDAtRPRWuVfxaM
4UgDg+PE6R/yUDGxPRbzxKgRiwLRqAr5CRcQ2WMnRvQ5sbcvIYUL6cTnzIeD
2XlhdN9h96j8vIhU+6xwxLq4ShpeX8kvrXlt0yu5AMSC+8Ia6N1b30Kapvgk
BDrn3Yv4ivsPU5AIC03vx1igMryGBMNSXEkfy5OC//bhrH9ynsiAOcQQdiCy
j1NgoDo1tt6v4ynYqOoBfw1fmX2Z9RXQ5ESGryqxn3RCbNdaFRXzA/LZ7X7f
0JYsMZW41D0FNxIhsJsi3YmStuph/uF1UVWwUV3REHePE4CRnz12B1zHp6eM
3/AqPbxf/F01GywVB2TzOKJpCyhxgkMmFs1s9iyqW9tNK6bj+eJNjW9IbWGY
9GhXPvadIDtkN/3j9hjYK6jssb7kkyVHRc0yofeYFmvzC4fAg7RBDCMWFQ==
END;
```

运行加密后的 plb 文件的方法如以下示例所示。

示例 14.3:运行被加密的文件,并查看加密的 PL/SQL 单元。在 KSQL 中,以运行加密文件 test1.plb 为例,创建函数 chksupply_totlwhamt。

程序代码如下。

选择子程序文本(被加密而不可读),然后调用子程序,运行并调用 test1.plb。代码如下。

```
seamart=# \i test1.plb
CREATE FUNCTION
seamart=# select prosrc from sys_proc where proname='chksupply_totlwhamt';
seamart-# /
          prosrc
-------------------------------------------------------------
 WRAPPED                                                    +
C810HCQWCGpd1PiSgX500ibAVLzwrBBTfuE18OOLy3dwEYItPSvYn43tmVj4+
 3srUFt5keky9LmUxCo41QWjSbB5SQW4raG+DxCZTIH5XxYo51GHUYJGb1GPP+
 7MFT8ujL2SHHKSI3Vl0w+8yL/7/6sQwBDsAHizQWuNObJ6/kDldL0ed27qNG+
 sFPUZ1U5kBM86tmxx7l2gP5DcSH+sd1ATQAAAAAAAAAAAAAAAAAAA=      +
 END
(1 行记录)
seamart=# DECLARE
seamart-#   RETURNVar INTEGER;
seamart-#   s_goodid BIGINT;
seamart-#   s_shopid BIGINT;
seamart-#   saleamt INTEGER;
seamart-# BEGIN
seamart-#   s_goodid :=12560557;
seamart-#   s_shopid :=104142;
seamart-#   saleamt :=2;
seamart-# RETURNVar:= "chksupply_totlwhamt"(s_goodid,s_shopid,saleamt);
seamart-#   RAISE NOTICE '%', RETURNVar;
```

```
seamart-# END;
seamart-# /
 998
```

14.3　DBMS_DDL 包的使用

14.3.1　使用 DBMS_DDL 子程序

DBMS_DDL 包提供了 WRAP 函数和 CREATE_WRAPPED 过程，二者都可以加密单个动态生成的可加密 PL/SQL 单元的源代码。如果 WRAP 或 CREATE_WRAPPED 输入的不是有效的可加密 PL/SQL 单元，则会引发异常。

每个 WRAP 函数将单个 CREATE 语句作为输入，该语句创建一个可加密的 PL/SQL 单元，并返回一个等效的 CREATE 语句，其中 PL/SQL 源代码已被加密。

每个 CREATE_WRAPPED 过程执行其对应的 WRAP 函数所完成的功能，然后运行返回的 CREATE 语句，创建指定的 PL/SQL 单元。

调用 DBMS_DDL 子程序时，使用完全限定的包名称 SYS.DBMS_DDL，以避免其他人再创建名为 DBMS_DDL 的本地包或定义公共同义词 DBMS_DDL 时发生名称冲突。

输入到 WRAP 函数或 CREATE_WRAPPED 过程的 CREATE 语句可以调用子程序的用户权限运行。

14.3.2　DBMS_DDL 加密的局限性

如果调用 DBMS_SQL.PARSE 解析 DBMS_DDL.WRAP 的输出（当为超过 32767B 的文本定义 VARCHAR2A 或者 VARCHAR2S 类型时），需要将 LFFLG 参数设置为 FALSE。否则，DBMS_SQL.PARSE 添加新行到加密单元中，并破坏程序单元。

14.3.3　示例

示例 14.4：使用 CREATE_WRAPPED 过程动态创建包规范（使用 EXECUTE IMMEDIAT 语句）和加密的包体。

程序代码如下。

```
DECLARE
  pkg_text   VARCHAR2(32767);                              -- 创建包规范和包体的文本
  FUNCTION generate_spec (pkgname VARCHAR2) RETURN VARCHAR2 AS
  BEGIN
    RETURN 'CREATE PACKAGE ' || pkgname || ' AUTHID CURRENT_USER AS
      PROCEDURE raise_price (good_id NUMBER, amount MONEY);
      END ' || pkgname || ';';
  END generate_spec;
  FUNCTION generate_body (pkgname VARCHAR2) RETURN VARCHAR2 AS
  BEGIN
    RETURN 'CREATE PACKAGE BODY ' || pkgname || ' AS
```

```
      PROCEDURE raise_price (good_id NUMBER, amount MONEY) IS
      BEGIN
         UPDATE goods SET price = price + amount WHERE goodid = good_id;
      END raise_price;
    END ' || pkgname || ';';
  END generate_body;
BEGIN
  pkg_text := generate_spec('good_prices');          --生成包规范文本
  EXECUTE IMMEDIATE pkg_text;                        --创建包规范
  pkg_text := generate_body('good_prices');          --生成包体文本
  SYS.DBMS_DDL.CREATE_WRAPPED(pkg_text);             --创建加密包体
END;
```

示例 14.5：选择示例 14.4 创建的包文本 good_prices，然后调用过程 good_prices.raise_price，如果包体被加密，那么调用过程所需的信息将不可读。

程序代码如下。

选择包文本的程序代码如下。

```
SELECT pkgbodysrc FROM sys_package WHERE pkgname = 'good_prices';
```

程序运行结果如下。

```
            pkgbodysrc
---------------------------------------------------------------
WRAPPED                                                       +
sS7LP7io7BJ2UEXlOlhSo0HIIrYrCHL3IE6U7gYU5eGMACwaixq/NwDZ9pZl+
fP5121Sl1zUs99jLD2fiOkpAsjtWGvnihfTzebAeKgX8rDo6ONGOk2sohRxJ+
VD52h4ecCDWsGlTEH62AM3CC3aQ9AgeIPPkuE7ftJWkA5WkashX/jnF5/AsQ+
vt5jX/2p/R/boV5jqwpzNzjrwAyf2AV4wSJSRf+tuk+VCtPuUlqDzSIXyzip+
Z983Ri/r1ou7L833                                              +
END
```

查询与调用的运行结果如下。

```
\set SQLTERM ;
SELECT * FROM goods WHERE goodid ='12';
goodid |  goodname  | price
------ +----------+------+
12     |    键盘    |$900.00 |
(3 rows)
CALL good_prices.raise_price(12, 2);
SELECT * FROM goods WHERE goodid ='12';
goodid |  goodname  | price
------ +----------+------+
12     |    键盘    |$900.00 |
12     |    键盘    |$902.00 |
(3 rows)
```

第 15 章

PL/SQL 的调试

通常,PL/SQL 都具备调试功能,以应对功能复杂情况下出现数据异常而需要分析的困难。KingbaseES 的调试器(debugger)能够让开发人员看到每一步骤的调试结果,准确地找到出错语句并修改。本章包括以下主要内容。

- PL/SQL 的执行跟踪。
- PL/SQL 调试器。

 15.1　PL/SQL 的执行跟踪

程序编译完成之后,会组织一些测试用例进行测试。如果测试结果出错,可以一头扎进源代码调试器中(总的来说,所有的 PL/SQL 编辑器提供了可视化的调试器,可以通过界面设置断点和观察点)查看。然而,首选应考虑跟踪程序的执行过程。

探讨 PL/SQL 代码跟踪之前,首先应区分调试和跟踪的不同之处。通常,首先需要对执行过程进行跟踪,获得应用程序行为的深度信息,以缩小问题源头;然后再通过调试器找到导致 bug 所在的具体代码行。

跟踪和调试的主要区别在于,跟踪是一个批量过程,而调试是交互过程。一个调试会话通常很耗时且枯燥,因此,任何能够把调试花费的时间最小化的工作都极具意义。主动跟踪可以做到这点。

每个应用程序都带有程序员自定义的跟踪机制,本节将探讨跟踪的选项。首先回顾实施跟踪时应该遵循的一些原则。

(1) 在开发部署各个阶段,对于跟踪的调用都要保留在代码中。不要在开发时插入跟踪调用,而部署到生产系统之前又去掉这些跟踪。当生产环境中真实用户使用该程序时,跟踪是开发人员了解程序运行过程的最好方法。

(2) 把调用跟踪功能的额外开销保持到最小。跟踪被禁用后,用户不应察觉到对应用系统性能有什么影响。

(3) 不要把在代码中直接调用 DBMS_OUTPUT.PUT_LINE 作为跟踪机制。对一些高质量要求的跟踪来说,这个内置功能不够灵活或强大。

(4) 调用 DBMS_UTILJTY.FORMAT_CALL_STACK 函数,以节省跟踪信息所使用的调用堆栈。

(5) 对最终用户而言,很容易启用或停用后端的跟踪代码,不需要技术支持就可以切换跟踪。同样也无须把有跟踪功能和无跟踪功能的应用做成两个版本。

(6) 如果其他人已经创建了一个同样可以使用的跟踪工具(也符合这里列出的以及自己定义的原则),则不必再创建一个自己的跟踪机制。

目前,KingbaseES 提供的主要跟踪工具如下。

(1) DBMS_UTILITY。

插件 dbms_utility 是 KingbaseES 的一个扩展插件。它的功能是提供 dbms_utility 系统包。dbms_utility 系统包提供一些具有通用功能的子程序,如查看错误堆栈、查看时间等。

(2) plsql_plprofiler。

插件 plsql_plprofiler 是 KingbaseES 的一个扩展插件。plsql_plprofiler 插件支持收集分析器(性能)数据,以提高性能或确定 PL/SQL 应用程序的代码覆盖率。应用程序可以使用代码覆盖率数据来集中增量测试工作。

此插件可以为所有 PL/SQL 单元生成分析信息,此信息包括每行的执行次数、执行该行花费的总时间,以及该行在多次执行中消耗的最长时间。

15.1.1 DBMS_UTILITY

1. DBMS_UTILITY 系统包子程序

DBMS_UTILITY 系统包包含的子程序及其简介如表 15.1 所示。

表 15.1　DBMS_UTILITY 系统包的子程序及其简介

子　程　序	简　　介
FORMAT_CALL_STACK 函数	返回当前的调用堆栈
FORMAT_ERROR_STACK 函数	返回当前的错误堆栈
FORMAT_ERROR_BACKTRACE 函数	返回当前的错误回溯信息
GET_TIME 函数	返回一个时间戳

2. FORMAT _CALL_STACK 函数

DBMS_UTILITY.FORMAT _CALL_STACK 函数可以返回当前的调用堆栈信息,包括调用对象的地址、调用的行号和调用对象的名称,可以用在函数、存储过程或触发器等 PL/SQL 对象中,用于帮助调试 PL/SQL 程序。

语法格式如下。

```
DBMS_UTILITY.FORMAT_CALL_STACK() RETURN TEXT;
DBMS_UTILITY.FORMAT_CALL_STACK(FORMAT TEXT) RETURN TEXT;
```

注意:目前调用对象的地址为函数或过程的 oid 值,如果是内建函数或过程,则为对应包的 oid 值。

说明如下。

FORMAT:指定调用堆栈信息的显示模式,分别有"o""p""s"三种模式。默认使用模式为"p"的模式,输出 8 位有效长度的十进制数代表对象的地址信息。

(1) 其中"o"参数将按照"%8x%8d%s"格式打印对象地址、调用行号、对象名称,如示例 15.2 所示。

(2) 其中"p"参数将按照"%8d%8d%s"格式打印对象地址、调用行号、对象名称,如示例 15.3 所示。

(3) 其中"s"参数将按照"%d,%d,%s"格式打印对象地址、调用行号、对象名称,如示例 15.4 所示。

示例 15.1

程序代码如下。

```
CREATE OR REPLACE FUNCTION checkCallStack() RETURNS TEXT AS
stack TEXT;
BEGIN
    SELECT dbms_utility.format_call_stack() INTO stack ;
    RETURN stack;
END;
call checkCallStack();
checkcallstack
------------------------------------------------------------
----- PL/SQL Call Stack ----- +
object line object +
handle number name +
14575 30 package body sys.dbms_utility.format_call_stack+
16387 4 function public.checkcallstack
(1 row)
```

示例 15.2

程序代码如下。

```
CREATE OR REPLACE FUNCTION checkHexCallStack() RETURNS TEXT AS
stack TEXT;
BEGIN
    SELECT dbms_utility.format_call_stack('o') INTO stack ;
    RETURN stack;
END;
CALL checkHexCallStack();
checkhexcallstack
------------------------------------------------------------
----- PL/SQL Call Stack ----- +
object line object +
handle number name +
38ef 22 package body sys.dbms_utility.format_call_stack+
4004 4 function public.checkhexcallstack
(1 row)
```

示例 15.3

程序代码如下。

```
CREATE OR REPLACE FUNCTION checkIntCallStack() RETURNS TEXT AS
stack TEXT;
BEGIN
    SELECT dbms_utility.format_call_stack('p') INTO stack ;
    RETURN stack;
END;
CALL checkIntCallStack();
checkintcallstack
----------------------------------------------------------------+
14575 22 package body sys.dbms_utility.format_call_stack+
16389 4 function public.checkintcallstack
(1 row)
```

示例 15.4

程序代码如下。

```
CREATE OR REPLACE FUNCTION checkIntUnpaddedCallStack() RETURNS TEXT AS
stack TEXT;
BEGIN
    SELECT dbms_utility.format_call_stack('s') INTO stack ;
    RETURN stack;
END;
call checkIntUnpaddedCallStack();
checkintunpaddedcallstack
---------------------------------------------+
14575,22,sys.dbms_utility.format_call_stack+
16390,4,public.checkintunpaddedcallstack
(1 row)
```

3. FORMAT_ERROR_BACKTRACE 函数和 FORMAT_ERROR_STACK 函数

（1）FORMAT_ERROR_BACKTRACE 函数用于返回当前程序发生异常时的错误回溯堆栈，若程序未发生异常，则返回 NULL。

语法格式如下。

```
DBMS_UTILITY.FORMAT_ERROR_BACKTRACE() RETURN TEXT;
```

（2）FORMAT_ERROR_STACK 函数用于返回当前程序发生异常时的错误堆栈，若程序未发生异常，则返回 NULL。

示例 15.5：format_error_backtrace 函数和 format_error_stack 函数在异常时使用。

程序代码如下。

```
CREATE OR REPLACE PROCEDURE p01() AS
    i INT := 0;
```

```
BEGIN
  i = i/0;
END;
CREATE OR REPLACE PROCEDURE p02() AS
  detail TEXT;
  stack TEXT;
BEGIN
    p01();
EXCEPTION
WHEN DIVISION_BY_ZERO THEN
    detail = dbms_utility.format_error_stack();
    stack = dbms_utility.format_error_backtrace();
    RAISE NOTICE 'FORMAT_ERROR_STACK IS: % ', detail;
    RAISE NOTICE 'FORMAT_ERROR_BACKTRACE IS: % ', stack;
END;
CALL p02();
```

程序运行结果如下。

```
FORMAT_ERROR_STACK IS: 除以零
at "sales.p01", line 4
FORMAT_ERROR_BACKTRACE IS: at "sales.p01", line 4
at "sales.p02", line 5
```

4. GET_TIME 函数

该函数用于返回一个时间戳,它不是标准的系统时间戳,单位为厘秒(百分之一秒)。通常在 PL/SQL 程序开始和结束时各调用一次该函数,然后用后一个数字减去前一个数字,以确定当前程序的执行耗时。

语法格式如下。

```
DBMS_UTILITY.GET_TIME() RETURN NUMBER;
```

说明如下。

函数的返回值为 $-2147483648 \sim 2147483647$,具体数值取决于机器和系统性能。调用该函数时,应用程序需要考虑返回值的符号。例如,两次调用皆为负数,或者第一次调用为负数、第二次调用为正数等情况。

示例 15.6:get_time 函数可用于计算语句的运行时间。

程序代码如下。

```
CREATE OR REPLACE PROCEDURE TestGetTime() AS
  t1   PLS_INTEGER;
  t2   PLS_INTEGER;
BEGIN
    t1 := dbms_utility.get_time();
    perform pg_sleep(3);
    t2 := dbms_utility.get_time();
```

```
        RAISE NOTICE 'sleeped: % sec.', (t2 - t1) / 100;
END;
call TestGetTime();
```

程序运行结果如下。

```
sleeped: 3 sec.
```

15.1.2 性能监控

1. 使用插件 plsql_plprofiler 进行性能监控

（1）插件加载。

KingbaseES 数据库默认将其添加到 kingbase.conf 文件的 shared_preload_libraries 中，重启数据库时自动加载。因此只需要在客户端工具执行 create extension 命令即可。程序代码如下。

```
-- 创建插件
create extension plsql_plprofiler;
```

（2）plsql_plprofiler 相关系统参数。

① plsql_plprofiler.max_callgraphs。

设置内存中存储堆栈关系的行数，默认值为 20000。

② plsql_plprofiler.max_functions。

设置内存中可存储的对象个数，默认值为 2000。

③ plsql_plprofiler.max_lines。

设置内存中分析数据的记录行数，默认值为 200000。

（3）plsql_plprofiler 相关函数。

plsql_plprofiler 中的监听方式分为本地监控、全局监控及指定会话监控 3 种。其中，本地监控模式是指只对当前 session 运行的对象进行监控；全局监控模式是指对所有 session 运行的对象进行监控；指定会话模式是指对某个指定 session 上运行的对象进行监控。

在 3 种模式同时启动的情况下，优先级为：全局监控模式＞指定会话监听模式＞本地监听模式。

plsql_plprofiler 中的数据共享方式分为 local 和 global 两种。其中，local 模式数据不共享，仅当前 session 可见；global 模式数据共享，所有 session 可见。相关函数和对应的功能如表 15.2 所示。

表 15.2 plsql_plprofiler 相关函数及其功能描述

函　　数	功　　能
pl_profiler_set_enabled_local 函数	启停当前 session 分析器
pl_profiler_set_enabled_pid 函数	启停指定 session 分析器
pl_profiler_get_enabled_local 函数	查看当前 session 分析器状态

续表

函 数	功 能
pl_profiler_get_enabled_pid 函数	查看指定 session 分析器状态
pl_profiler_set_enabled_global 函数	启停数据库级分析器
pl_profiler_get_enabled_global 函数	查看全局监控是否启用的状态
pl_profiler_collect_data 函数	单次数据 flush
pl_profiler_set_collect_interval 函数	指定时间间隔数据 flush
pl_profiler_get_collect_interval 函数	查看定时迁移的状态
pl_profiler_reset_local 函数	本地数据 clean
pl_profiler_reset_shared 函数	全局数据 clean
pl_profiler_linestats_local 函数	查看本地的对象的执行信息
pl_profiler_linestats_shared 函数	查看全局的对象的执行信息
pl_profiler_func_oids_local 函数	查看本地已执行 PL/SQL 对象
pl_profiler_func_oids_shared 函数	查看全局已执行 PL/SQL 对象
pl_profiler_callgraph_local 函数	查看本地的堆栈关系
pl_profiler_callgraph_shared 函数	查看全局的堆栈关系
pl_profiler_get_stack 函数	OID 转对象名
pl_profiler_funcs_source 函数	查看指定对象源码信息
pl_profiler_version 函数	查看整型版本信息
pl_profiler_versionstr 函数	查看字符串版本信息
pl_profiler_lines_overflow 函数	查看行数是否过界

（4）plsql_plprofiler 使用。

支持 plsql_plprofiler 的对象如下。

① 支持函数分析。

② 支持存储过程分析。

③ 支持包分析。

包中支持分析的对象包括包中初始化块、包中函数、包中存储过程。包中对象的分析数据记录到包对应的 oid。

④ 支持触发器分析。

⑤ 支持嵌套函数分析。

⑥ 支持匿名块分析。

匿名块的分析数据记录在行号为 9999 的 oid 中。

⑦ 支持 object type 分析。

object type 中支持分析的对象包括 member 方法、construct 方法、static 方法。object type 中分析数据记录到各个方法对应的 oid 中。

注意：如果重复执行同一个对象（以 oid 为标识），需要清理之前记录的数据。

支持 plsql_plprofiler 的数据信息分为 3 类：PL/SQL 对象执行信息、PL/SQL 对象堆栈信息和 PL/SQL 对象源码信息。

(5) plsql_plprofiler 获取的执行信息。

① 对象 oid。

plsql_plprofiler 对象的 oid，包中的所有对象共用包的 oid，匿名块的 oid 为 9999。object type 中的对象使用方法本身的 oid，嵌套函数的 oid 使用父函数的 oid。

② 语句所在行。

0 行表示该对象的整体数据。

如果 PL/SQL 源码在一行上，则该语句的任何执行相关数据都属于该行。

如果 PL/SQL 源码在多行上，则该语句的任何执行相关数据都属于该语句的第一行。

如果 PL/SQL 源码的一行出现多个语句时，则多条语句的任何执行相关数据都属于该行。

③ 执行次数。

记录某一行代码的执行次数。一行代码不一定是一条语句。

④ 总共耗时。

记录某一行代码的总耗时。如果是控制语句，则包括整个控制语句中包含的语句时间，时间单位为微秒（μs）。

⑤ 最长耗时。

记录某一行代码耗时最长的时间，时间单位为 μs。

注意：

1. 当执行对象抛出的异常未被捕获时，只会记录整个对象的执行时间，其他数据均为 0。当执行对象抛出的异常被捕获时，异常语句的数据为 0，其他语句数据正常记录。

2. 如果执行对象中有事务操作（commit/rollback），该对象记录的总耗时是事务操作前的总时间。

3. 当开启 set serverout on 时，每条 SQL 语句都会执行 dbms_output 包中的函数。因此执行语句时，plsql_plprofiler 会记录调用 dbms_output 包的数据，如调用数据 clean 函数后，首先清除执行数据，然后调用 dbms_ouput 包中的函数。

(6) plsql_plprofiler 获取的堆栈信息。

① 堆栈。

plsql_plprofiler 对象的堆栈。

② 调用次数。

该对象被调用的总次数。

③ 总耗时。

该对象总花费时间。

④ 调用对象耗时。

该对象调用对象的耗时。

⑤ 自身耗时。

该对象除去调用其他对象的耗时。

(7) plsql_plprofiler 获取的源码信息。

① oid。

② 行号。

③ 源码。

由于匿名块在数据库中没有被存储，所以无法获取匿名块的源码信息。

(8) plsql_plprofiler 的运行步骤。

plsql_plprofiler 每次运行的过程包括以下步骤。

① 启动分析器。

② 运行应用程序。

③ 关闭分析器。

④ 数据 flush。

⑤ 分析数据。

使用 plsql_plprofiler 前，需要创建 plsql_plprofiler 扩展。

plsql_plprofiler 收集数据，通过调用 pl_profiler_set_enabled_local 函数来控制分析器的启动或停止。分析器收集的数据存储在内存中，如果断开连接，分析器将不会自动存储数据，需要显示调用 pl_profiler_collect_data 函数将数据刷到共享内存。需要注意的是，目前 plsql_plprofiler 数据不会落盘存储，所以一旦重启数据库，分析数据将会丢失。

示例 15.7：使用编辑查询调用 plsql_plprofiler 插件里的函数，给出每个语句的执行时间和执行次数，从而诊断性能问题。

程序代码如下。

```
--创建扩展
create extension plsql_plprofiler;
--创建对象
create table t1(i int);

create or replace procedure p1() as
begin
    raise notice 'this is procedure p1';
    for i in 1..1000 loop
        insert
        into
        t1
        values(99999);
    end loop;
    if 1000 < 100 then null; else null; end if;
end;

create or replace procedure p2() as
begin
    raise notice 'this is procedure p2';
    p1();
end;
```

```
create or replace procedure p3() as
begin
    raise notice 'this is procedure p3';
    p2();
    if 1 < 100 then
        raise notice '1 < 100';
    end if;
end;
```

--清理本地数据
```
select pl_profiler_reset_local();
```
--清理全局数据
```
select pl_profiler_reset_shared();
```
--启动分析器
```
select pl_profiler_set_enabled_global(true);
```
--运行应用程序
```
call p3();
```
--关闭分析器
```
select pl_profiler_set_enabled_global(false);
```
--数据 flush
```
select pl_profiler_collect_data();
```
--分析数据
----查看已记录的对象(可选)
```
select * from pl_profiler_func_oids_shared();
    Name                        |Value             |
--------------------------------+------------------+
pl_profiler_func_oids_shared    |{26540,26542,26541} |
```
----oid转对象名(可选)
```
select * from pl_profiler_get_stack(pl_profiler_func_oids_shared());
pl_profiler_get_stack                                          |
---------------------------------------------------------------+
{sales.p1() oid=26540,sales.p3() oid=26542,sales.p2() oid=26541}|
```
----查看数据(可选)
```
Select func_oid, func_oid::regproc as funcname, line_number, source from pl_profiler_funcs_source(pl_profiler_func_oids_shared());
func_oid |funcname |line_number |source                                          |
---------+---------+------------+------------------------------------------------+
  26540  |p1       |     0      |-- Line 0                                       |
  26540  |p1       |     1      |                                                |
  26540  |p1       |     2      |begin                                           |
  26540  |p1       |     3      |    raise notice 'this is procedure p1';        |
  26540  |p1       |     4      |    for i in 1..1000 loop                       |
  26540  |p1       |     5      |        insert                                  |
  26540  |p1       |     6      |        into                                    |
  26540  |p1       |     7      |        t1                                      |
  26540  |p1       |     8      |        values(99999);                          |
  26540  |p1       |     9      |    end loop;                                   |
  26540  |p1       |    10      |    if 1000 < 100 then null; else null; end if; |
  26540  |p1       |    11      |end                                             |
  26542  |p3       |     0      |-- Line 0                                       |
```

```
 26542   |p3       |         1 |                                                                    |
 26542   |p3       |         2 |begin                                                               |
 26542   |p3       |         3 |  raise notice 'this is procedure p3';                              |
 26542   |p3       |         4 |  p2();                                                             |
 26542   |p3       |         5 |  if 1 < 100 then                                                   |
 26542   |p3       |         6 |    raise notice '1 < 100';                                         |
 26542   |p3       |         7 |  end if;                                                           |
 26542   |p3       |         8 |end                                                                 |
 26541   |p2       |         0 |-- Line 0                                                           |
 26541   |p2       |         1 |                                                                    |
 26541   |p2       |         2 |begin                                                               |
 26541   |p2       |         3 |  raise notice 'this is procedure p2';                              |
 26541   |p2       |         4 |  p1();                                                             |
 26541   |p2       |         5 |end                                                                 |
```

----查看对象调用关系

select * **from** pl_profiler_callgraph_shared();

```
stack                  |call_count |us_total |us_children |us_self |
-----------------------+-----------+---------+------------+--------+
{26542,26541}          |         1 |   14305 |      13726 |    579 |
{26542}                |         1 |   69855 |      14305 |  55550 |
{26542,26541,26540}    |         1 |   13726 |          0 |  13726 |
```

-----源码和分析数据对应该系

SELECT L.func_oid::**regproc as** funcname,
    L.func_oid **as** func_oid,
    L.line_number,
    sum(L.exec_count)::**bigint AS** exec_count,
    sum(L.total_time)::**bigint AS** total_time,
    max(L.longest_time)::**bigint AS** longest_time,
    S.source
FROM pl_profiler_linestats_shared() L
JOIN pl_profiler_funcs_source(pl_profiler_func_oids_shared) S
    ON S.func_oid = L.func_oid **AND** S.line_number = L.line_number
    GROUP BY L.func_oid, L.line_number, S.source
ORDER BY L.func_oid, L.line_number;

```
funcname |func_oid |line_number |exec_count |total_time |longest_time |source                                            |
---------+---------+------------+-----------+-----------+-------------+--------------------------------------------------+
p1       | 26540   |0           |1          |13726      |13726        |-- Line 0                                         |
p1       | 26540   |1           |0          |0          |0            |                                                  |
p1       | 26540   |2           |1          |13722      |13722        |begin                                             |
p1       | 26540   |3           |1          |185        |185          |raise notice 'this is procedure p1';              |
p1       | 26540   |4           |1          |13367      |13367        |for i in 1..1000 loop                             |
p1       | 26540   |5           |1000       |12564      |1543         |insert                                            |
p1       | 26540   |6           |0          |0          |0            |into                                              |
p1       | 26540   |7           |0          |0          |0            |t1                                                |
p1       | 26540   |8           |0          |0          |0            |values(99999);                                    |
p1       | 26540   |9           |0          |0          |0            |end loop;                                         |
p1       | 26540   |10          |2          |0          |0            |if 1000 < 100 then null; else null; end if;       |
p1       | 26540   |11          |0          |0          |0            |end                                               |
p2       | 26541   |0           |1          |14305      |14305        |-- Line 0                                         |
p2       | 26541   |1           |0          |0          |0            |                                                  |
p2       | 26541   |2           |1          |14304      |14304        |begin                                             |
```

p2	26541	3	1	261	261	raise notice 'this is procedure p2';
p2	26541	4	1	14037	14037	p1();
p2	26541	5	0	0	0	end
p3	26542	0	1	69855	69855	-- Line 0
p3	26542	1	0	0	0	
p3	26542	2	1	69853	69853	begin
p3	26542	3	1	42675	42675	raise notice 'this is procedure p3';
p3	26542	4	1	26962	26962	p2();
p3	26542	5	1	211	211	if 1 < 100 then
p3	26542	6	1	156	156	raise notice '1 < 100';
p3	26542	7	0	0	0	end if;
p3	26542	8	0	0	0	end

2. 使用 KStudio 图形化工具进行性能监控

示例 15.8：通过 KStudio 图形化工具获取每行代码的执行时间和执行次数，进而诊断性能问题。

在 KStudio 中通过 CALL 关键字调用并执行函数和存储过程，可以单击 SQL 编辑器最上方的性能报告，查看 SQL 语句的执行信息，如图 15.1 所示。

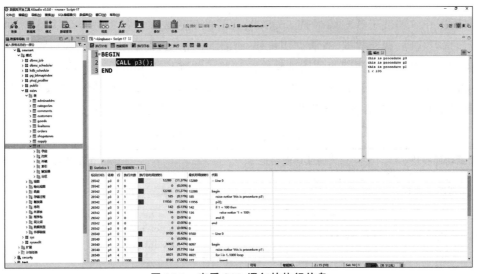

图 15.1　查看 SQL 语句的执行信息

15.2　PL/SQL 调试器

调试器（debugger）主要针对存储过程、函数、触发器、程序包进行调试。本节使用插件中的函数和 KStudio 数据库开发工具进行调试。

15.2.1　函数/存储过程调试

1. 插件 pldbgapi 和 plsql_pldbgapi

（1）pldbgapi。

插件 pldbgapi 是 KingbaseES 的一个扩展插件，主要用于 PL/pgSQL 语言的存储过程/

函数的代码调试。使用时加载该插件,不需要时卸载插件即可。

(2) plsql_pldbgapi。

插件 plsql_pldbgapi 是 KingbaseES 的一个扩展插件。主要用于 PL/SQL 语言的存储过程、函数、包、触发器函数、匿名块、object type 方法的代码调试。使用时加载该插件,不需要时卸载插件即可。plsql_pldbgapi 拓展插件不支持 PG 模式。

以下介绍插件 pldbgapi 的使用,plsql_pldbgapi 与 pldbgapi 的使用方法相同,只需要替换插件里的存储过程即可。

(1) 加载插件。

在 KingbaseES V8R6 之前,使用 pldbgapi 时,需要将它添加到 kingbase.conf 文件的 shared_preload_libraries 中,并重启 KingbaseES 数据库。该版本及之后的版本中不需要此操作,plugin_debugger 已经预定义到 conf 文件中。程序代码如下。

```
shared_preload_libraries = 'plugin_debugger'
-- 创建插件
create extension pldbgapi;
```

(2) pldebugger 相关函数。

pldebugger 调试工具支持对 PL/pgSQL 语言的函数或存储过程进行调试操作,包括操作断点、执行代码、调试数据、查看调用堆栈、查看源码信息、查看版本信息等功能,并提供局部断点调试与全局断点调试两种调试模式。这些函数都位于 pldbgapi 扩展下,使用前需要创建 pldbgapi 扩展。pldebugger 针对 PL/pgSQL 提供了多个调试模块,每个模块提供了相应的接口函数,以支持各种调试操作。支持的功能模块和功能特性如表 15.3 所示。

表 15.3 pldebugger 相关函数及其功能特性

调试功能模块	调试功能特性	调试功能函数
建立局部断点流程	创建局部监听 连接目标后端	plpgsql_oid_debug pldbg_attach_to_port
建立全局断点流程	创建全局监听 设置全局断点 等待连接	pldbg_create_listener pldbg_set_global_breakpoint pldbg_wait_for_target
操作断点	设置断点 删除断点 返回建立连接时断点所在行的信息	pldbg_set_breakpoint pldbg_drop_breakpoint pldbg_wait_for_breakpoint
执行代码	单步进入(step into) 单步跳过(step over) 继续执行(continue) 中止执行	pldbg_step_into pldbg_step_over pldbg_continue pldbg_abort_target
调试数据	查看变量 修改变量	pldbg_get_variables pldbg_deposit_value
调用堆栈	查看堆栈信息 追溯堆栈信息	pldbg_get_stack pldbg_select_frame
查看信息	查询源代码信息 查询当前系统的版本信息	pldbg_get_source pldbg_get_proxy_info

(3) pldebugger 调试流程。

以下详细介绍局部断点调试的命令行调试流程。全局断点调试流程只是在目标后端与调试端建立连接的流程方面有细微差别,实际的调试流程总体一致。pldebugger 的调试操作是建立在目标后端和调试端正常通信的前提下,其详细的调试流程如下。

① 创建监听:局部断点调试根据调试对象的 OID 建立监听,全局断点调试则不需要带调试对象的 OID,而是创建全局监听。

② 创建连接:全局调试与局部调试的连接方式稍有不同。对于局部调试,首先调用待调试的对象,让目标后端进程以通信服务器端的身份发起连接,然后在调试端调用端口连接函数 pldbg_attah_to_port,让调试端以通信客户端的身份与目标后端建立连接;对于全局断点调试,先设置全局断点,然后调用 pldbg_wait_for_target 函数发起连接,并等待接受目标后端的连接。此时调用待调试的对象就可以与调试端建立通信连接。

③ 调试设置:可以根据需要设置和删除断点信息,包括局部断点和全局断点。

④ 执行调试操作:pldebugger 提供了 3 种调试执行的方式,分别为单步进入(step into)、单步跳过(step over)以及继续执行(continue)。

⑤ 信息查询:在调试执行的过程中,可以查询源代码信息、查询和修改变量值、查询调用堆栈以及当前系统版本等。

示例 15.9:创建调试环境和用例。

注:本例只需要开启 2 个 SQL 编辑器(一个用作应用端,一个用作调试端)即可。这里为更好地说明调试过程,帮助读者更好地理解调试顺序,而将其分开操作。

程序代码如下。

```
--创建 pldebugger 环境
create extension pldbgapi;
--创建用来调试的存储过程
CREATE OR REPLACE PROCEDURE public.pro1()
language 'plpgsql'
AS $$
BEGIN
  raise notice '123';
  raise notice 'success';
END;
$$;

CREATE OR REPLACE PROCEDURE public.pro4()
LANGUAGE 'plpgsql'
AS
$$
declare
  i int := 2;
begin
  i = 3;
  raise notice 'i= %',i;
  i = 4;
  raise notice 'i= %',i;
```

```
    raise notice '0:ready.....';
  begin
    raise notice '1:ready.....';
    call pro1();
    begin
      raise notice 'success';
    end;
  end;
end;
$$;
--查询出需要调试的存储过程的 oid
select oid from sys_proc where proname = 'pro4';
    oid  |
------- +
16648|
```

应用端程序代码如下。

```
--应用端开启局部监听
select plpgsql_oid_debug(16648);
    plpgsql_oid_debug |
----------------- +
                 0 |
--开始调试存储过程 pro4
call pro4();
==============================
PLDBGBREAK:11 --该进程控制器转交给另一进程,直至调试结束,11 为连接的端口号
```

调试端程序代码如下:

```
--调试端连接
select pldbg_attach_to_port(11);
pldbg_attach_to_port   |
-------------------- +
                   1|
--查看源码
  select pldbg_get_source(1,16648);
pldbg_get_source
---------------------------------
declare
  i int := 2;
begin
  i = 3;
  raise notice 'i= %',i;
  i = 4;
  raise notice 'i= %',i;
  raise notice '0:ready.....';
  begin
    raise notice '1:ready.....';
```

```
      call pro1();
      begin
        raise notice 'success';
      end;
    end;
end;
```
--设置断点
```
select pldbg_set_breakpoint(1,16648,7);
    pldbg_set_breakpoint |
--------------------+
true                     |
select pldbg_set_breakpoint(1,16648,12);
pldbg_set_breakpoint     |
--------------------+
true                     |
select pldbg_set_breakpoint(1,16648,15);
    pldbg_set_breakpoint |
--------------------+
true                     |
```
--获取当前设置的断点信息
```
select pldbg_get_breakpoints(1);
pldbg_get_breakpoints    |
--------------------+
(16648,7,"")             |
(16648,12,"")            |
(16648,-1,"")            |
(16648,15,"")            |
```
--单步执行,遇到子程序进入
```
select pldbg_step_into(1);
    pldbg_step_into      |
--------------------+
(16648,7,"pro4()")       |
select pldbg_step_into(1);
pldbg_step_into          |
--------------------+
(16648,8,"pro4()")       |

select pldbg_step_into(1);
pldbg_step_into          |
--------------------+
(16648,9,"pro4()")       |
```
--单步执行,遇到子程序跳过
```
select pldbg_step_over(1);
pldbg_step_over          |
--------------------+
(16648,10,"pro4()")      |
```
--继续执行程序,直到断点处或者程序结束
```
select pldbg_continue(1);
pldbg_continue           |
--------------------+
```

```
 (16648,13,"pro4()")
--当application client执行完毕后,debug client进程会挂起,直到application
client再次执行测试函数或存储过程
select pldbg_continue(1);
```

应用端程序代码如下。

```
--应用端存储过程执行完成
call pro4();
PLDBGBREAK:11
i= 3
i= 4
0:ready.....
1:ready.....
123
success
success
--再次调起
call pro4();
```

调试端应用程序代码如下。

```
--调试端进入调试状态,调用pldbg_abort_target,终止调试
select pldbg_abort_target(1);
pldbg_abort_target   |
-------------------+
true               |
```

应用端程序代码如下。

```
--当debug client调用select pldbg_abort_target(1)后,执行被终止
call pro4();
  SQL 错误 [57014]:错误:由于用户请求而正在取消查询
  Where:在赋值的第5行的PL/pgSQL函数pro4()
```

2. 使用 KStudio 工具进行调试

下面以函数 chksupply_totlwhamt 为例,说明 PL/SQL 的函数/存储过程调试过程。

(1) 在左侧导航栏中找到 seamart 数据库 sales 模式下的"函数"节点,右击函数名 chksupply_totlwhamt,在弹出的菜单中选择"调试"命令,弹出"函数参数"对话框,如图 15.2 所示,输入参数,进入图 15.3 所示的"调试 PL/SQL"窗口。

(2) 单击左上角的调试按钮 开始调试,选择工具栏中的"单步跳过""单步步入""单步返回""继续""暂挂""终止"等按钮,继续调试程序,直到完全正确为止,如图 15.4 所示。

(3) 可以在调试窗口的右上侧窗口区域查看导入断点,以便查看程序运行到此处时是否发生错误,如图 15.5 所示。

第15章 PL/SQL 的调试

图 15.2 "函数参数"对话框

图 15.3 PL/SQL 函数调试

（4）可以在调试窗口的右上侧窗口区域查看函数执行过程变量的变化以及最后的结果，如图 15.6 所示。

15.2.2 触发器调试

下面以示例 13.1 创建 DML 触发器为例，说明触发器的调试过程。

（1）在左侧导航栏中找到 seamart 数据库 sales 模式下的"触发器"节点，右击触发器名 tri，在弹出的菜单中选择"调试"命令，弹出"触发器调试"对话框，如图 15.7 所示，选择触发器操作的表、触发类型，输入相关的参数，进入图 15.8 所示的"金仓调试 PL/SQL"窗口。

图 15.4 调试的具体步骤、结果(一)

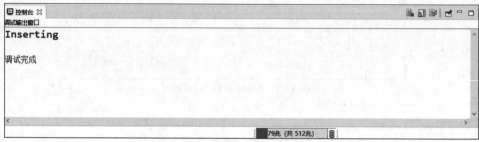

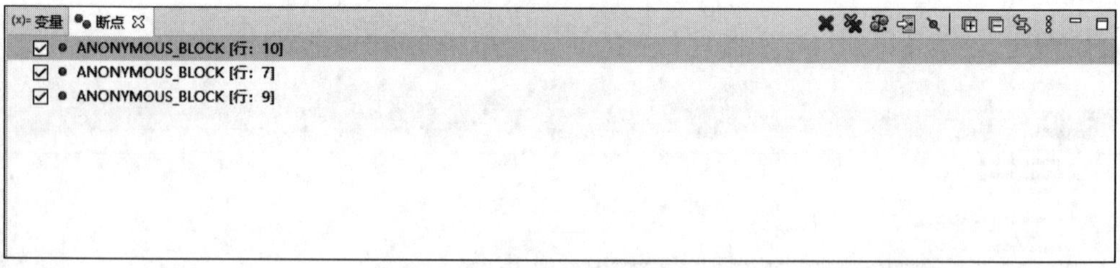

图 15.5 设置、查看断点(一)

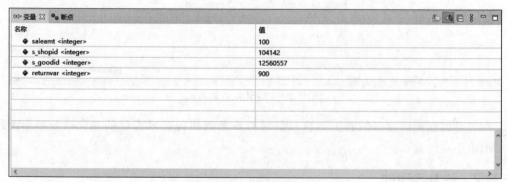

图 15.6 变量状态(一)

(2) 输入必要的参数后(如像主键之类的非空约束必须写,否则在调试过程中报错),单击左上角的调试按钮" "开始调试,选择工具栏中的"单步跳过""单步步入""单步返回""继续""暂挂""终止"等按钮,继续调试程序,直到调试完全正确为止,如图 15.9 所示。

图 15.7 "触发器调试"对话框

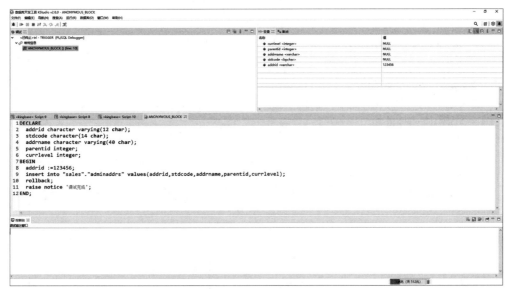

图 15.8 触发器调试(一)

(3)可以在调试窗口的代码区域设置断点,在要添加断点的行双击行最左侧,即可添加断点,然后在右上侧窗口区域导入断点,以便查看触发器运行到此处时是否发生错误,如图 15.10 所示。

(4)可以在调试窗口的右上侧窗口区域查看触发器执行过程变量的变化以及最后的结果,如图 15.11 所示。

15.2.3 程序包调试

下面以示例 12.2 创建包 emp_good 为例,说明程序包的调试过程。

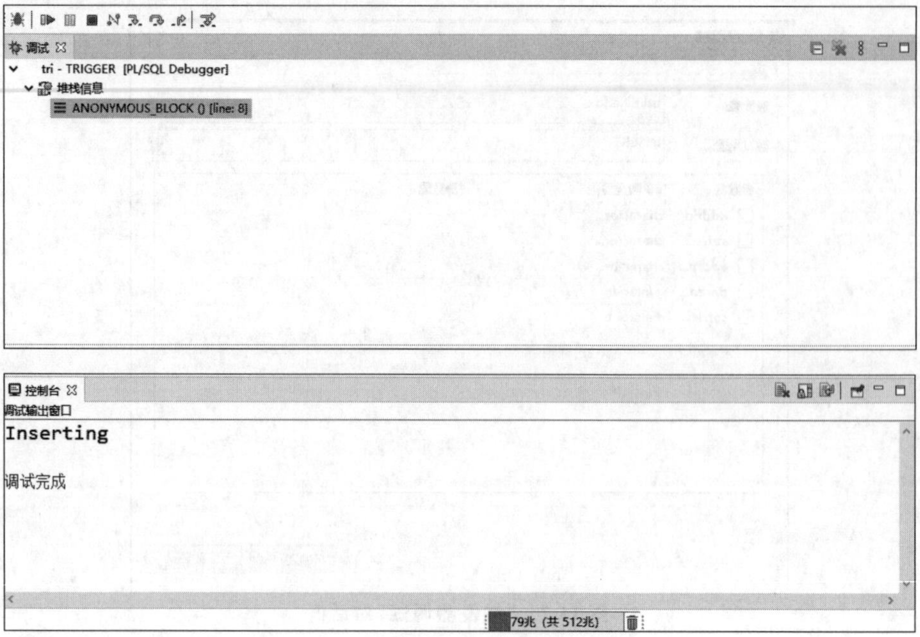

图 15.9　调试的具体步骤、结果（二）

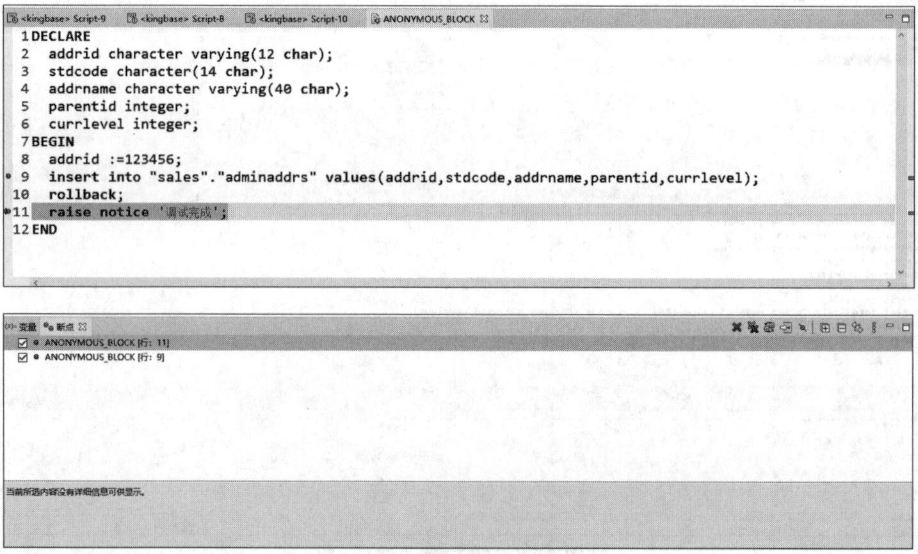

图 15.10　设置、查看断点（二）

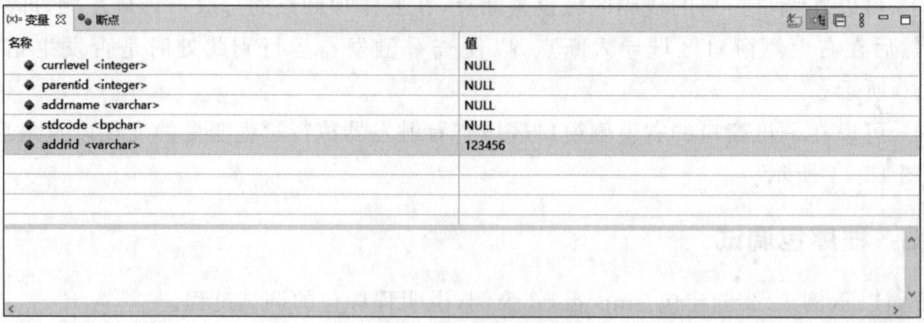

图 15.11　变量状态（二）

（1）在左侧导航栏中找到 seamart 数据库 sales 模式下的"程序包"节点，右击程序包 emp_good，在弹出的菜单中选择"调试"命令，弹出"程序包参数"对话框，如图 15.12 所示，选择包内部函数和存储过程，输入相关参数，进入图 15.13 所示的"金仓调试 PL/SQL"窗口。

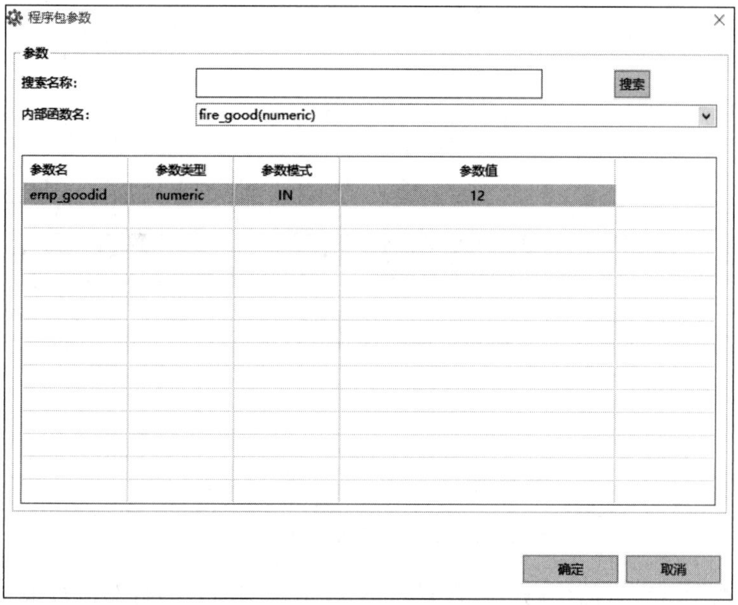

图 15.12　"程序包参数"对话框

图 15.13　触发器调试（二）

（2）输入必需的参数后，单击左上角调试按钮" "开始调试，选择工具栏中的"单步跳过""单步步入""单步返回""继续""暂挂""终止"等按钮，继续调试程序，直到调试完全正确为止，如图 15.14 所示。

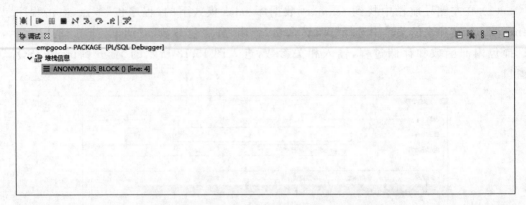

图 15.14 调试的具体步骤、结果(三)

(3)可以在调试窗口的代码区域设置断点,在要添加断点的行双击行最左侧,即可添加断点,然后在右上侧窗口区域导入断点,以便查看包中函数或存储过程运行到此处时是否发生错误,如图 15.15 所示。

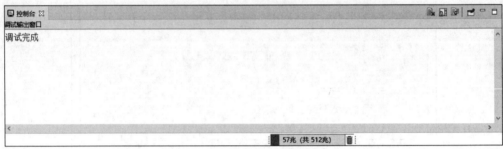

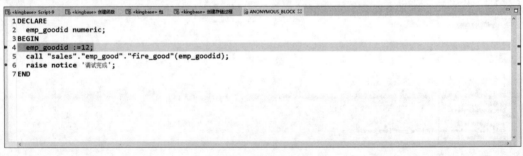

图 15.15 设置、查看断点(三)

(4)可以在调试窗口的右上侧窗口区域查看包中函数或存储过程所执行过程变量的变化,最后的结果如图 15.16 所示。

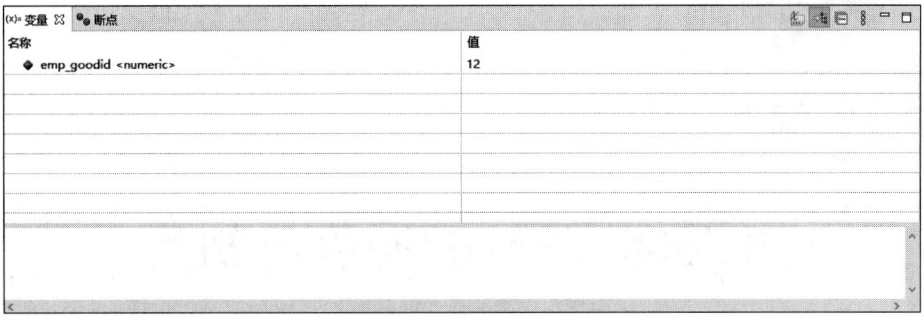

图 15.16　变量状态（三）

第 16 章

PL/SQL 任务的调度与执行

KingbaseES 在使用任务调度前需要在配置文件中添加 shared_preload_libraries = 'kdb_schedule', job_queue_processes=1000。KingbaseES 基于后台进程实现自动作业功能。通过 kdb_schedule 插件提供 DBMS_JOB 和 DBMS_SCHEDULER 包,其中定义了自动作业功能的相关函数。本章包括以下主要内容。

- 使用 DBMS_JOB 包管理任务。
- 使用 DBMS_SCHEDULER 包管理任务。
- 使用 KStudio 管理任务。

 ## 16.1 使用 DBMS_JOB 包管理任务

DBMS_JOB 模式提供调度和管理作业任务的功能,可以由 DBMS_SCHDULER 替代。DBMS_JOB 具有如下限制。

(1)需要创建 kdb_schedule 扩展。
(2)仅支持 PL/SQL 类型的任务。
(3)仅支持本地任务。
(4)间隔时间采用日历表示法。

表 16.1 列出了 DBMS_JOB 各子程序及其概要描述。

表 16.1 DBMS_JOB 软件包子程序

类 别	子 程 序	描 述
任务的创建	SUBMIT Procedure	创建一个 job
	WHAT Procedure	更改 job 执行的动作
	CHANGE Procedure	更改 job 的运行参数
	INTERVAL Procedure	更改 job 运行的时间
	NEXT_DATE Procedure	更改 job 下一次的运行时间
任务的运行与终止	BROKEN Procedure	终止一个处于可执行状态的 job
	RUN Procedure	立即运行指定的 job
任务的删除	REMOVE Procedure	删除一个 job

16.1.1 任务的创建

1. 使用 SUBMIT 存储过程创建一个任务

使用 DBMS_JOB 包中的 submit 可以直接创建任务,此过程创建一个 job,并且返回一个由系统分配的 job 的 ID,以便后面使用。

语法格式如下。

```
DBMS_JOB.SUBMIT(
    OUT job INTEGER,
    what text,
    next_date TIMESTAMP DEFAULT now(),
    interva text DEFAULT NULL,
    no_parse BOOLEAN DEFAULT FALSE,
    instance INTEGER DEFAULT 0,
    force BOOLEAN DEFAULT FALSE);
```

参数说明如表 16.2 所示。

表 16.2 SUBMIT 参数

参　　数	描　　述
job	由系统分配的 job 的 ID
what	job 要运行的 PL/SQL
next_date	job 运行的下一个日期
interva	job 运行的间隔时间
no_parse	暂不支持,默认为 false
instance	暂不支持,默认为 0
force	暂不支持,默认为 false

示例 16.1

程序代码如下。

```
call dbms_job.submit(1, 'CALL "sales"."updatesupply_totlwhamt"(12560557,
104142,1)', now(), 'Freq=daily;BYHOUR=23;BYMINUTE=0;BYSECOND=0', false, 0,
false);
```

程序运行结果如下。

```
Name |Value |
----+----+
job  |23   |
```

2. INSTANCE 存储过程

此过程增加 job 执行的连接串信息,即数据库连接的一些基本信息,参数依赖于创建任务时系统分配的 jobid，意味着必须先创建任务，拿到 jobid 才能正确执行该代码片段。

语法格式如下。

```
DBMS_JOB.INSTANCE(
    job INTEGER,
    instance TEXT);
```

参数说明如表 16.3 所示。

表 16.3　INSTANCE 参数

参　　数	描　　述
job	由系统分配的 job 的 ID
instance	job 要运行数据库的连接串

示例 16.2

程序代码如下。

```
CALL dbms_job.instance(23, 'user=system dbname=seamart port=54321 password=123456');
--查询任务
SELECT * FROM "kdb_schedule"."kdb_action";
```

程序运行结果如下。

```
Name           |Value                                                        |
---------------+-------------------------------------------------------------+
acid           |15                                                           |
acname         |                                                             |
acdesc         |                                                             |
acenabled      |false                                                        |
ackind         |s                                                            |
accode         |CALL "sales"."updatesupply_totlwhamt"(12560557,104142,1)     |
acconnstr      |                                                             |
acdbname       | seamart                                                     |
acnextrun      |                                                             |
```

查询任务的程序代码如下。

```
SELECT * FROM "kdb_schedule"."kdb_schedule_job";
```

程序运行结果如下。

```
Name           |Value |
---------------+------+
sjid           |20    |
sjscid         |14    |
sjjobid        |23    |
```

```
sjstatus    |s    |
sjlasttime  |     |
sjnexttime  |2022-12-03 15:24:35.644736+08|
```

3. 改变任务的属性

以下存储过程用来在 DBMS_JOB 包中修改已经创建好的任务属性。这些存储过程创建好任务后，以系统为任务分配的 jobid 作为参数，因而，在此之前必须有对应的创建任务语句执行。

（1）CHANGE 存储过程。

此过程更改 job 中的属性信息。

语法格式如下。

```
DBMS_JOB.CHANGE(
    job INTEGER,
    what TEXT,
    next_date TIMESTAMP,
    interva TEXT,
    instance INTEGER DEFAULT 0,
    force BOOLEAN DEFAULT FALSE);
```

参数说明如表 16.4 所示。

表 16.4 CHANGE 参数

参数	描述
job	由系统分配的 job 的 ID
what	job 要运行的 PL/SQL
next_date	job 运行的下一个日期
interva	job 运行的间隔时间
instance	暂不支持，默认为 0
force	暂不支持，默认为 false

示例 16.3

程序代码如下。

```
--改变执行的 plsql 语句
CALL dbms_job.change(23, 'CREATE TABLE T1(a INT)', now(), 'Freq=daily;BYHOUR=23;BYMINUTE=0;BYSECOND=0', 0, false);
--查询任务
SELECT * FROM "kdb_schedule"."kdb_action";
```

程序运行结果如下。

```
Name            |Value                           |
--------------- +--------------------------------+
acid            |15                              |
acname          |                                |
acdesc          |                                |
acenabled       |false                           |
ackind          |s                               |
accode          | CREATE TABLE T1(a INT)         |
acconnstr       |                                |
acdbname        | seamart                        |
acnextrun       |                                |
```

(2) INTERVAL 存储过程。

此过程更改 job 中的间隔时间属性。

语法格式如下。

```
DBMS_JOB.INTERVAL(job INTEGER,interva TEXT);
```

参数说明如表 16.5 所示。

表 16.5 INTERVAL 参数

参数	描述
job	由系统分配的 job 的 ID
interval	job 运行的间隔时间

示例 16.4

程序代码如下。

```
CALL dbms_job.interval(23,'Freq=daily;BYHOUR=10;BYMINUTE=10;
BYSECOND=10');
--查询任务
SELECT * FROM "kdb_schedule"."kdb_schedule";
```

程序运行结果如下。

```
Name                  |Value                                            |
--------------------- +-------------------------------------------------+
scid                  |14                                               |
scname                |                                                 |
scdesc                |                                                 |
scenabled             |true                                             |
scstart               |2022-12-03 15:24:35.644736                       |
scend                 |                                                 |
screpeat_interval     | Freq=daily;BYHOUR=10;BYMINUTE=10;BYSECOND=10    |
```

(3) NEXT_DATE 存储过程。

此过程更改 job 中的下次运行时间信息。

语法格式如下。

```
DBMS_JOB.NEXT_DATE(
    job INTEGER,
    next_date TIMESTAMP);
```

参数说明如表 16.6 所示。

表 16.6 NEXT_DATE 参数

参　　数	描　　述
job	由系统分配的 job 的 ID
next_date	job 运行的下一个日期

示例 16.5

程序代码如下。

```
CALL dbms_job.next_date(23, now());
--任务查询
SELECT * FROM "kdb_schedule"."kdb_schedule_job";
```

程序运行结果如下。

```
Name        |Value                        |
------------+-----+
sjid        |20                           |
sjscid      |14                           |
sjjobid     |23                           |
sjstatus    |s                            |
sjlasttime  |                             |
sjnexttime  |2022-12-03 15:24:35.644736+08|
```

（4）WHAT 存储过程。

此过程更改 job 执行的行为信息。

语法格式如下。

```
DBMS_JOB.WHAT(
    job INTEGER,
    what TEXT);
```

参数说明如表 16.7 所示。

表 16.7 WHAT 参数

参　　数	描　　述
job	由系统分配的 job 的 ID
what	job 要运行的 PL/SQL

示例 16.6

程序代码如下。

```
CALL dbms_job.what(23, 'CALL updatesupply_totlwhamt(12560557,104142,1)');
--任务查询
SELECT * FROM "kdb_schedule"."kdb_action";
```

程序运行结果如下。

```
Name            |Value
----------------+------------------------------------------------------
acid            |15
acname          |
acdesc          |
acenabled       |false
ackind          |s
accode          |CALL updatesupply_totlwhamt(12560557,104142,1)
acconnstr       |
acdbname        | seamart
acnextrun       |
```

16.1.2 任务的执行

上一小节介绍了任务创建语句,本小节主要介绍任务的执行与中止。

1. RUN 存储过程

此过程运行一个 job。

语法格式如下。

```
CALL DBMS_JOB.RUN(job INTEGER);
```

参数说明如表 16.8 所示。

表 16.8 RUN 参数

参　　数	描　　述
job	由系统分配的 job 的 ID

示例 16.7

程序代码如下。

```
--调用 dbms_job.run();启动 Job
CALL dbms_job.run(23);
```

2. BROKEN 存储过程

此程序将 job 设置为终止,而不再运行。

语法格式如下。

```
DBMS_JOB.BROKEN(
    job INTEGER,
    broken BOOLEAN,
    next_date TIMESTAMP DEFAULT now());
```

参数说明如表 16.9 所示。

表 16.9 BROKEN 参数

参数	描述
job	由系统分配的 job 的 ID
broken	将 job 设置为 enable 或 disable。true 为 disable，false 为 enable
next_date	job 运行的下一个日期

示例 16.8

程序代码如下。

```
--调用 dbms_job.broken();中止 Job
CALL dbms_job.broken(23,TRUE,now());
```

16.1.3 任务的删除

DBMS_JOB 的 REMOVE 方法用来删除一个 job。
语法格式如下。

```
DBMS_JOB.REMOVE(job INTEGER);
```

参数说明如表 16.10 所示。

表 16.10 REMOVE 参数

参数	描述
job	由系统分配的 job 的 ID

示例 16.9

程序代码如下。

```
CALL dbms_job.remove(23);
--任务查询
SELECT * FROM "kdb_schedule"."kdb_schedule_job";
```

程序运行结果如下。

```
Name       |Value |
-----------+------+
sjid       |      |
sjscid     |      |
```

sjjobid		
sjstatus		
sjlasttime		
sjnexttime		

示例 16.10：综合运用 DBMS_JOB 包创建 1 个备份数据的任务，每分钟执行 1 次。程序代码如下。

```
CREATE OR REPLACE PROCEDURE sales.copy_shopstores_2023_01_05()
AS
BEGIN
  DROP TABLE IF EXISTS sales.shopstores_temp ;
  CREATE TABLE sales.shopstores_temp AS (SELECT * FROM sales.shopstores);
END

DECLARE
  jobid int;
BEGIN
  call dbms_job.submit(jobid,'call sales.copy_shopstores_2023_01_05()',now(),
'Freq=minutely;interval=1;bysecond=10', false, 0, false);
  call dbms_job.instance(jobid, ' user = system dbname = seamart port = 54321 password=123456');
  call dbms_job.run(jobid);
  RAISE NOTICE '%',jobid;
END;

SELECT * FROM "kdb_schedule"."kdb_joblog" kj;
```

程序运行结果如下。

```
jlgid |jlgjobid |jlgstatus |jlgstart           |jlgduration                                |
----+--------+---------+-----------------+------------------------------+
 20  |50      |s        |2023-01-07 20:14:03.657608+08 | 0 years 0 mons 0 days 0 hours 0 mins 0.527021 secs |
 21  |49      |f        |2023-01-07 20:14:13.709898+08 | 0 years 0 mons 0 days 0 hours 0 mins 1.486667 secs |
 22  |50      |s        |2023-01-07 20:14:14.182729+08 |0 years 0 mons 0 days 0 hours 0 mins 1.43020 secs |
 23  |50      |s        |2023-01-07 20:15:14.273692+08 |0 years 0 mons 0 days 0 hours 0 mins 1.18979 secs |
 24  |49      |f        |2023-01-07 20:15:23.867048+08 | 0 years 0 mons 0 days 0 hours 0 mins 0.642653 secs |
 25  |50      |s        |2023-01-07 20:16:13.933199+08 | 0 years 0 mons 0 days 0 hours 0 mins 0.558406 secs |
 26  |50      |s        | 2023-01-07 20:17:13.94323+08 | 0 years 0 mons 0 days 0 hours 0 mins 0.563859 secs |
 27  |49      |f        |2023-01-07 20:17:33.929318+08 | 0 years 0 mons 0 days 0 hours 0 mins 0.509020 secs |
```

```
28     |50   |s   |2023-01-07 20:18:13.989017+08   | 0 years 0 mons 0 days 0 hours 0 mins 0.528470
secs   |
```

16.2 使用 DBMS_SCHEDULER 包管理任务

DBMS_SCHEDULER 模式提供调度和管理计划任务的功能。此包包含了 DBMS_JOB 包的功能，并有所增强。

DBMS_SCHEDULER 具有如下限制。

（1）需要创建 kdb_schedule 扩展。

（2）间隔时间采用日历表示法。

表 16.11 列出了 DBMS_SCHEDULRE 各子程序及其概要描述。

表 16.11 DBMS_SCHEDULER 软件包子程序

类 别	子 程 序	描 述
任务的创建	CREATE_PROGRAM Procedure	创建一个程序
	CREATE_SCHEDULE Procedure	创建一个计划
	CREATE_JOB Procedure	创建一个计划任务
任务的删除	DROP_JOB Procedure	删除一个计划任务
	DROP_PROGRAM Procedure	删除一个程序
	DROP_SCHEDULE Procedure	删除一个计划
任务的执行	RUN_JOB Procedure	运行一个计划任务

16.2.1 任务的创建

使用 DBMS_SCHEDULER 包中的子程序创建任务时，必须先创建程序和计划才可以创建一个具体的任务。原因在于，当创建任务时，会用到已经创建好的程序名和计划名。

1. CREATE_PROGRAM Procedure

创建一个 job 任务的程序。

程序对象描述调度器要运行的内容。程序是一个独立于作业的实体，作业则在某个确定的时间运行，并调用某个程序。可以创建指向现有程序对象的作业，这意味着不同的作业可以使用相同的程序，并在不同的时间以不同的设置运行该程序。有了正确的权限，不同的用户可以使用相同的程序，而不必重新定义它。因此，创建程序库后，用户可以从现有程序列表中加以选择。

语法格式如下。

```
DBMS_SCHEDULER.CREATE_PROGRAM(
program_name TEXT,
program_type TEXT,
```

```
program_action TEXT,
acconnstr TEXT,
acdbname TEXT,
number_of_arguments INTEGER DEFAULT 0,
enabled BOOLEAN DEFAULT FALSE,
comments TEXT DEFAULT NULL
);
```

参数说明如表 16.12 所示。

表 16.12 CREATE_PROGRAM 参数

参数	描述
program_name	程序的名字
program_type	程序的类型。有下列类型：PLSQL_BLOCK、STORED_PROCEDURE、SQL_SCRIPT、EXECUTABLE、EXTERNAL_SCRIPT、BACKUP_SCRIPT
program_action	程序的动作
acconnstr	数据库连接串
acdbname	数据库名称
number_of_arguments	程序动作的参数，暂不支持，0 为默认值
enabled	程序的状态，true 为启动状态，false 为禁用状态
comments	程序的注释信息

需要注意的是，PLSQL_BLOCK 必须大写，acdbname 必须指定。

acdbname 指明了 action 所在的数据库，如果没有指定，默认为 kdb_schedule 运行时指定的数据库。

示例 16.11

程序代码如下。

```
--创建 program
BEGIN
  dbms_scheduler.create_program(
    program_name       => 'prog_02',
    program_type       => 'PLSQL_BLOCK',
    program_action     => 'CALL "sales"."updatesupply_totlwhamt"(12560557,
104142,1)',
    acconnstr          => 'user=system dbname=seamart port=54321 password=123456',
    acdbname           => 'seamart',
    number_of_arguments => 0,
    enabled            => true,
    comments           => 'test program');
END;
--成功创建 program 后，会有如下一行信息：
SELECT * FROM kdb_schedule.kdb_action;
```

```
Name        |Value                                                      |
------------+-----------------------------------------------------------+
acid        |18                                                         |
acname      |prog_02                                                    |
acdesc      |test program                                               |
acenabled   |true                                                       |
ackind      |s                                                          |
accode      |CALL updatesupply_totlwhamt(12560557,104142,1)             |
acconnstr   |user=system dbname=seamart port=54321 password=123456      |
acdbname    |seamart                                                    |
acnextrun   |                                                           |
```

2. CREATE_SCHEDULE Procedure

创建一个 job 任务的调度程序。

调度计划对象（schedule）指定作业何时运行以及运行多少次。调度计划可以被多个作业共享。例如，对于许多作业来说，一个商业季度的结束可能是一个常见的时间框架。与每次季度末定义新作业不同，作业创建者可以指向一个已命名的调度计划。

KingbaseES 支持基于时间的调度，可以让作业立即运行或稍后运行。时间调度包括开始日期及时间、结束日期及时间、重复间隔。

语法格式如下。

```
DBMS_SCHEDULER.CREATE_SCHEDULE(
schedule_name TEXT,
start_date TIMESTAMP WITH TIME ZONE DEFAULT NULL,
repeat_interval TEXT DEFAULT NULL,
end_date TIMESTAMP WITH TIME ZONE DEFAULT NULL,
comments TEXT DEFAULT NULL
);
```

参数说明如表 16.13 所示。

表 16.13　CREATE_SCHEDULE 参数

参　　数	描　　述
schedule_name	调度程序的名字
start_date	调度程序的开始时间
repeat_interval	调度程序的间隔时间
end_date	调度程序的结束时间
comment	调度程序的注释信息

示例 16.12

程序代码如下。

```
--创建 Schedule
BEGIN
    dbms_scheduler.create_schedule(
```

```
    schedule_name   => 'schedule_02',
    start_date      => now(),
    repeat_interval => 'freq=minutely;interval=1',
    end_date        => null,
    comments        => 'test schedule');
END;
--创建 schedule 后,会有如下一行信息:
SELECT * FROM kdb_schedule.kdb_schedule;
Name                   |Value                                   |
-----------------------+----------------------------------------+
scid                   |17                                      |
scname                 |schedule_02                             |
scdesc                 |test schedule                           |
scenabled              |true                                    |
scstart                |2022-12-03 17:42:16.967539              |
scend                  |                                        |
screpeat_interval      |freq=minutely;interval=1                |
```

3. CREATE_JOB Procedure

创建一个 job 任务。

若完成了程序和调度计划的创建,意味着可以基于两者创建一个有计划的作业(job)。语法格式如下。

```
DBMS_SCHEDULER.CREATE_JOB(
job_name TEXT,
program_name TEXT,
schedule_name TEXT,
job_class TEXT DEFAULT 'Routine Maintenance',
enabled BOOLEAN DEFAULT FALSE,
auto_drop BOOLEAN DEFAULT TRUE,
comments TEXT DEFAULT NULL,
credentail_name TEXT DEFAULT NULL,
destination_name TEXT DEFAULT NULL
);
```

参数说明如表 16.14 所示。

表 16.14 CREATE_JOB 参数

参数	描述
job_name	任务的名字
program_name	程序的名字
schedule_name	计划的名字
job_class	任务的类型,默认为 Routine Maintenance
enable	任务的状态。true 为启用状态,false 为禁用状态
auto_drop	任务完成后自动删除,暂不支持,默认为 true

续表

参　数	描　述
comment	任务的注释信息
credentail_name	暂不支持,默认为 NULL
destination_name	暂不支持,默认为 NULL

示例 16.13

程序代码如下。

```
--创建 Job
BEGIN
    dbms_scheduler.create_job(
    job_name        => 'job_02',
    program_name    => 'prog_02',
    schedule_name   => 'schedule_02',
    job_class       => 'routine maintenance',
    enabled         => true,
    auto_drop       => true,
    comments        => 'test job',
    credentail_name => null,
    destination_name => NULL
);
END;
--创建后,下表将有相关记录信息,分别记录 job 与 schedule,以及 job 与 action 之间的关系。
SELECT * FROM kdb_schedule.kdb_schedule_job;
Name          |Value |
---------+-----+
sjid          |23    |
sjscid        |17    |
sjjobid       |24    |
sjstatus      |s     |
sjlasttime    |      |
sjnexttime    |      |
SELECT * FROM kdb_schedule.kdb_job_action;
Name          |Value |
---------+-----+
jaid          |23    |
jajobid       |24    |
jaacid        |18    |
jastatus      |s     |
jalasttime    |      |
--具体的 job 信息,可以看 kdb_job:
SELECT jobid,jobname,jobenabled,joblastrun,jobnextrun,jobrepeattimes FROM kdb_job;
Name               |Value                     |
------------+--------------------+
jobid              |24                        |
jobname            |job_02                    |
```

```
jobenabled      |true                          |
joblastrun      |                              |
jobnextrun      |2022-12-03 17:42:16+08        |
jobrepeattimes  |0                             |
```

16.2.2 任务的执行

RUN_JOB Procedure 用来运行 job 任务。通过调用 DBMS_SCHEDULER.RUN_JOB,可以使用 RUN_JOB 程序来测试作业或在调度计划之外运行它。RUN_JOB 中的 use_current_session 用来确定作业是异步还是同步,默认值为 true。

语法格式如下。

```
DBMS_SCHEDULER.RUN_JOB(
job_name TEXT,
use_current_session BOOLEAN DEFAULT TRUE
);
```

参数说明如表 16.15 所示。

表 16.15 RUN_JOB 参数

参　　数	描　　述
job_name	任务的名字
use_current_session	暂不支持,默认值为 true

示例 16.14

程序代码如下。

```
CALL dbms_scheduler.run_job('job_02', true);
```

16.2.3 任务的删除

使用 DBMS_SCHEDULER 包中的子程序删除任务时,必须先删除任务才能删除程序和计划。原因在于,任务中使用程序名和计划名,没有删除任务却删除程序和计划会导致报错。

1. DROP_JOB Procedure

删除一个 job 任务。

语法格式如下。

```
DBMS_SCHEDULER.DROP_JOB(
job_name TEXT,
force BOOLEAN DEFAULT FALSE,
defer BOOLEAN DEFAULT FALSE,
commit_semantics TEXT DEFAULT 'STOP_ON_FIRST_ERROR'
);
```

参数说明如表 16.16 所示。

表 16.16　DROP_JOB 参数

参　　数	描　　述
job_name	job 的名字
force	暂不支持,默认值为 false
defer	暂不支持,默认值为 false
commit_semantice	暂不支持,默认值为 STOP_ON_FIRST_ERROR

示例 16.15

程序代码如下。

```
-- 删除一个job任务
BEGIN
  dbms_scheduler.drop_job(
    job_name    => 'job_02'
  );
END;
```

2. DROP_SCHEDULE Procedure

删除一个 job 任务的调度程序,在此之前必须删除与调度程序相关的 job 任务。
语法格式如下。

```
DBMS_SCHEDULER.DROP_ SCHEDULE (
schedule_name TEXT,
force BOOLEAN DEFAULT FALSE
);
```

参数说明如表 16.17 所示。

表 16.17　DROP_ SCHEDULE 参数

参　　数	描　　述
schedule_name	调度程序的名字
force	暂不支持,默认值为 false

示例 16.16

程序代码如下。

```
-- 删除一个job任务的调度程序
BEGIN
  dbms_scheduler.drop_schedule(
    schedule_name    => 'schedule_02'
  );
END;
```

3. DROP_PROGRAM Procedure

删除一个 job 任务的程序,在此之前必须删除与程序相关的 job 任务。

语法格式如下。

```
DBMS_SCHEDULER.DROP_PROGRAM (
program_name TEXT,
force BOOLEAN DEFAULT FALSE
);
```

参数说明如表 16.18 所示。

表 16.18 DROP_PROGRAM 参数

参 数	描 述
program_name	程序的名字
force	暂不支持,默认值为 false

示例 16.17

程序代码如下。

```
-- 删除一个job任务的程序
BEGIN
  dbms_scheduler.drop_program(
    program_name    => 'prog_02'
  );
END;
```

示例 16.18:综合运用 DBMS_SCHEDULER 包创建 1 个备份数据的任务,每分钟执行 1 次。

程序代码如下。

```
BEGIN
  CALL DBMS_SCHEDULER.CREATE_PROGRAM('program5', 'PLSQL_BLOCK', 'call sales.copy_shopstores1()','user=system dbname=seamart port=54321 password=123456', 'test', 0, true, 'this is test program');
  CALL DBMS_SCHEDULER.CREATE_SCHEDULE('schedule5', now(), 'Freq=minutely;interval=1;bysecond=10', NULL, 'this is test schedule');
  CALL DBMS_SCHEDULER.CREATE_JOB('job5', 'program5', 'schedule5', 'Routine Maintenance', true, true, 'this is test job');
  CALL DBMS_SCHEDULER.RUN_JOB('job5', true);
END
SELECT * FROM "kdb_schedule"."kdb_joblog" kj;
```

程序运行结果如下。

```
jlgid |jlgjobid |jlgstatus |jlgstart              |jlgduration                      |
----  +-------- +--------- +--------------------- +-------------------------------- +
```

```
274    |58    |s    |2023-01-08 20:48:08.259695+08    |0 years 0 mons 0 days 0 hours 0 
mins 0.545914 secs    |
275    |58    |s    |2023-01-08 20:48:18.295246+08    |0 years 0 mons 0 days 0 hours 0 
mins 1.615370 secs    |
276    |50    |s    |2023-01-08 20:48:18.855824+08    |0 years 0 mons 0 days 0 hours 0 
mins 1.65053 secs    |
277    |58    |s    |2023-01-08 20:49:18.381533+08    |0 years 0 mons 0 days 0 hours 0 
mins 1.577849 secs    |
278    |50    |s    |2023-01-08 20:49:18.899683+08    |0 years 0 mons 0 days 0 hours 0 
mins 1.101930 secs    |
279    |56    |f    |2023-01-08 20:49:38.374822+08    |0 years 0 mons 0 days 0 hours 0 
mins 0.594637 secs    |
```

16.3　使用 KStudio 管理任务

16.3.1　任务的创建

1. 创建一个程序

在 KStudio 界面下单击所需创建任务的数据库，打开后再选择计划任务，再选择程序，右击，在弹出的对话框中选择新建程序，填写相应的信息，单击"确定"按钮即可创建一个程序。下面以创建数据库表 t3 任务为例，说明用图形化界面管理任务，如图 16.1 所示。

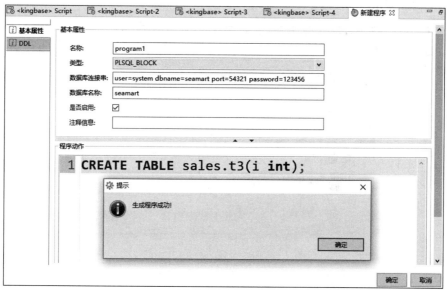

图 16.1　创建程序

2. 创建一个计划

在 KStudio 界面下单击所需创建任务的数据库，打开后再选择计划任务，然后选择计划，右击，在弹出的对话框中选择新建计划，填写相应的信息，单击"确定"按钮即可创建一个计划，如图 16.2 所示。

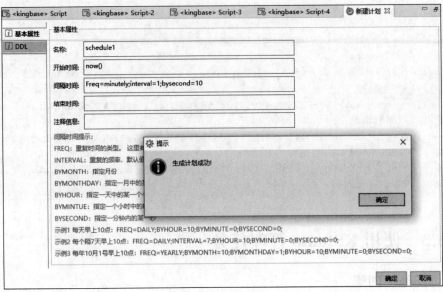

图 16.2　创建计划

3. 创建一个任务

在 KStudio 界面下单击所需创建任务的数据库，打开后再选择计划任务，然后选择任务，右击，在弹出的对话框中选择新建任务，填写相应的信息，单击"确定"按钮即可创建一个任务，如图 16.3 所示。

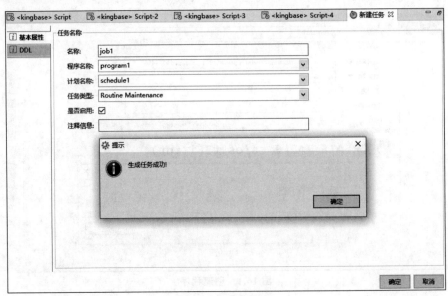

图 16.3　创建任务

16.3.2　任务的执行

在创建任务的目录下右击所要执行的任务，选择运行任务，弹出"执行计划任务"对话框，如图 16.4 所示，单击"确定"按钮后提示执行任务成功，如图 16.5 所示。

第 16 章 PL/SQL 任务的调度与执行

图 16.4 "执行计划任务"对话框

图 16.5 执行任务成功

执行结果如图 16.6 所示。

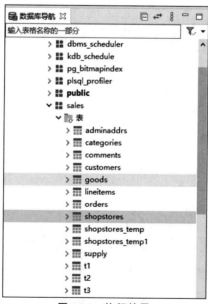

图 16.6 执行结果

16.3.3 任务的删除

1. 删除 job 任务

在创建任务的目录下选择要删除的任务,右击,在弹出的对话框中选择删除即可,如图 16.7 所示。

图 16.7 删除 job 任务

2. 删除计划

在创建任务的目录下选择要删除的计划,右击,在弹出的对话框中选择删除即可。需要注意的是,删除计划之前必须确保与之相关的任务已经删除,否则就会报错,如图 16.8 所示。

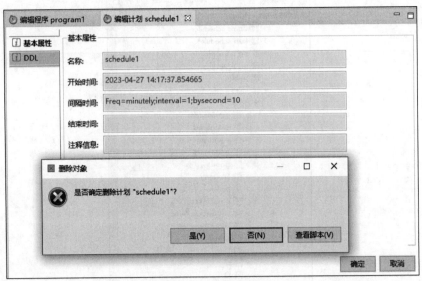

图 16.8 删除计划

3. 删除程序

在创建任务的目录下选择要删除的程序,右击,在弹出的对话框中选择删除即可。需要注意的是,删除程序之前必须确保与之相关的任务已经删除,否则就会报错,如图 16.9 所示。

图 16.9 删除程序

参考文献

[1] 北京人大金仓信息技术股份有限公司. KingbaseES 联机文档. V9[EB/OL]. [2023-03-06]. https://help.kingbase.com.cn/v8/index.html.

[2] 王珊,萨师煊. 数据库系统概论[M]. 5 版. 北京:高等教育出版社,2014.

[3] Silberschatz A,Korth H F, Sudarshan S. 数据库系统概念(原书第 6 版)[M]. 杨冬青,李红燕,唐世渭,译. 北京:机械工业出版社,2012.

[4] 彭煜玮. PostgreSQL 12.2 手册[M]. PostgreSQL 中文社区文档翻译组,译. [2023-02-10]. http://www.postgres.cn/docs/12/.

[5] 谭峰,张文升. PostgreSQL 实战[M]. 北京:机械工业出版社,2018.

[6] 屠要峰,陈河堆. 深入浅出 PostgreSQL[M]. 北京:电子工业出版社,2022.

[7] PriceJ. 精通 Oracle Database 12c SQL&PL/SQL 编程[M]. 3 版. 卢涛,译. 北京:清华大学出版社,2014.

[8] Feuerstein S,Priby B. Oracle PL/SQL 程序设计上下册[M]. 方鑫,译. 北京:人民邮电出版社,2022.

[9] 孙风栋,王澜,郭晓惠. Oracle 12c PL/SQL 程序设计终极指南[M]. 6 版. 北京:机械工业出版社,2015.

[10] ORACLE. PL/SQL Language Reference[EB/OL]. [2019-01-01]. https://docs.oracle.com/en/database/oracle/oracle-database/12.2/lnpls/index.html.

[11] ORACLE. SQL Language Reference[EB/OL]. [2019-01-01].https://docs.oracle.com/en/database/oracle/oracle-database/12.2/sqlrf/index.html.

[12] Morton K,Osborne K,Sands R,et al. 精通 Oracle SQL[M]. 2 版.朱浩波,译. 北京:人民邮电出版社,2014.

[13] Anon.ISO/IEC JTC 1/SC 32 Data management and interchange Technical Committee. Information technology:Database languages:SQL:Part 2: Foundation (SQL/Foundation)(ISO/IEC 9075-2:2016). [2016-12-01]. https://www.iso.org/standard/63556.html.

[14] Nanda A,Tierney B,Helskyaho H. SQL 和 PL/SQL 深度编程[M]. 唐波,侯圣文,译. 北京:清华大学出版社,2019.

[15] Burleson D K, Celko J, Cook J P, et al. Advanced SQL Database Programmers Handbook[M]. United States of America:BMC Software and DBAzine,2003.